Rodger Bradley

LNER 4-6-0s

DAVID & CHARLES
Newton Abbot   London   North Pomfret (VT)

British Library Cataloguing in Publication Data

Bradley, Rodger P., *1952–*
LNER 4-6-0s.
1. Great Britain. Railway services. London
and North Eastern Railway 4-6-0 steam
locomotives to 1960
I. Title
625.2'61'0941

ISBN 0–7153–8895–9

Photoset and printed in Great Britain by
Redwood Burn Limited, Trowbridge, Wiltshire
for David & Charles (Publishers) plc
Brunel House   Newton Abbot   Devon

Published in the United States of America
by David & Charles Inc
North Pomfret   Vermont 05053   USA

B12/3 No 61565, rebuilt by Gresley, is seen at
Nottingham Victoria in March 1953. *Roger Shenton*

Title page. While self-weighing tenders were not so
extensively used as their counterparts on the LMS, B1
No 61258 is here seen on a half-day excursion from
Lincoln to Dudley in July 1960, with one of these
possessed by the LNER. This locomotive was seen a
couple of months later having BR standard aws fitted
at Doncaster Works. *Roger Shenton*

Jacket. Preserved LNER Class B1 4-6-0 Mayflower.
Roger Siviter

Above right. The second LNER 4-6-0 to carry the name
City of London – the first having been Class B2 No 5427
– fitted with a streamlined casing and classified B17/5
for working the East Anglian service. *National Railway
Museum*

Right. GCR Class 9Q No 461, later to become LNER
Class B7, at Neasden in 1948, with its four cylinders all
in line under the smokebox. The heavy casting for the
slidebars support, and incorporating a footstep, is
clearly seen in this view. The locomotive was built by
Vulcan Foundry in 1921, and survived until August
1948. *L&GRP*

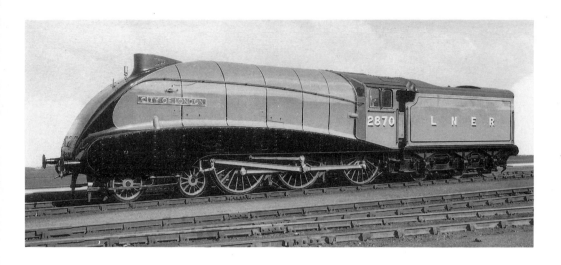

CONTENTS

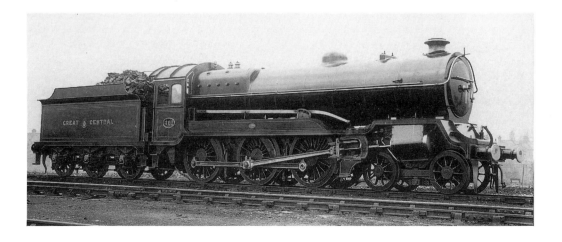

INTRODUCTION AND ACKNOWLEDGEMENTS

The London & North Eastern Railway possessed some 17 4-6-0 types, but only two of these were not pre-Grouping designs, and one was a rebuild of the single new design produced during Gresley's reign as chief mechanical engineer. The most successful – at least in terms of the number of locomotives built – was the Thompson B1 design, of which 409 were turned out between 1942 and 1950. Sir Nigel Gresley – who was in fact the second choice as cme of the newly-constituted company in 1923 – produced only one new type, the B17 Sandringham class 4-6-0, which was developed to replace or reinforce the lightweight 4-6-0s already in service on the Great Eastern lines.

The majority of 4-6-0 classes were provided by the North Eastern Railway and Great Central Railway companies, which handed over four and nine different types

respectively. The North East of England, which had effectively been the birthplace of railways, also gave birth to the first British passenger 4-6-0, following the successful adopting of this wheel arrangement by David Jones for freight work on the Highland Railway. It is interesting to reflect that a representative of this first passenger 4-6-0 actually survived in departmental use to become the property of the nationalised British Railways in 1948. The NER under Wilson Worsdell and Sir Vincent Raven produced a further three 4-6-0 types, which later became classes B14, B15 and B16 in the LNER classification. The last-mentioned survived well into BR days, and some rebuilding by both Gresley and Thompson, while those of the former Great Central Railway, although providing the greatest variety in design, fared much less well. John Robinson, the incumbent chief mechanical engineer of the GCR at the time of the Grouping, was invited to become the cme of the new company, but turned down the appointment as at 65 he was

Possibly the first visit of a B1 to the West Coast of Scotland, No 61344 is on the turntable at Oban, on 19 July 1949, still carrying the 'British Railways' legend on the tender. *Roger Shenton*

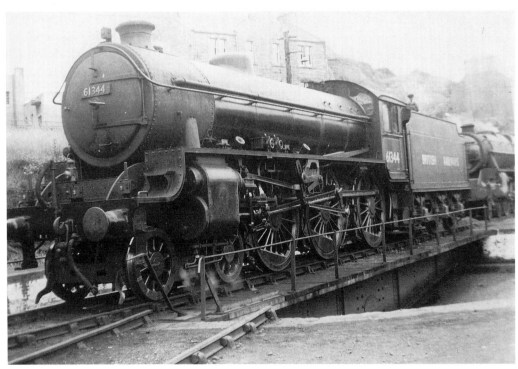

B2 4-6-0 No 5427 *City of London* still on express passenger duties. The locomotives were later reclassified B19, to make way for the Thompson rebuilds of Class B17. The name was subsequently carried by streamlined Class B17 No 2870 for the East Anglian service. *L&GRP*

near retiring age, and suggested the appointment of Gresley. Robinson left in his wake a number of 4-6-0 types, some of which had some serious design faults which were not easily remedied, while all seemed to acquire a degree of notoriety in their economy of operation. Still, the LNER was not an affluent company, although perhaps its expenditure on a small number of express passenger designs may have given the opposite impression, and it was considered better to continue with modification many of the existing secondary types. Reconstruction of some pre-Grouping designs ranged from the fitting of poppet valve gear, through major mechanical alterations, the installation of Gresley's favoured three-cylinder conjugated valve gear, to provision of new boilers. Later, even Gresley's designs received attention during years of austerity, leading to essentially simpler and more accessible layout of machinery.

Some comparative trials in the late 1940s with one of these modified 4-6-0 designs – the B17 – tended to indicate that it was not so much an improvement in the mechanics of the locomotive that was gained, but a more efficient and effective boiler design. The B17 was provided with a new boiler for the tests, such as was fitted to the new B1 class 4-6-0 by Edward Thompson, and it turned out to be more economical and efficient on its old chassis than on the new two-cylinder layout devised by Thompson in the B2 rebuild. The B1 with the same boiler was extensively tested during 1950 and 1951, and again proved the merits of the new boiler design, as the official results of these tests ably demonstrated. To many, this last LNER 4-6-0 appeared completely out of step with traditional practices established under Gresley, but it was in essence a synthesis of many of those same practices.

During BR days, particularly during the middle 1950s and later, only the B1, B17, B12 and B16 4-6-0s represented the type on the Eastern and North Eastern Regions. Obviously a major factor in the disappearance of the type from the rails was the advent of dieselisation, and most were withdrawn in the early 1960s. While it is true that examples of the LNER's 4-6-0s have been rescued for preservation – the B1 and B12 designs – it is perhaps regrettable that others have not survived.

I am indebted to a number of people in the preparation of this book. In particular I would like to thank Sid Checkley for his advice, Roger Shenton and Geoff Sharpe for a number of illustrations, the Gresley Society, Philip Atkins and the library staff in the National Railway Museum, and for invaluable help in assisting with research the staff of the Reference Section of the Harris Library in Preston, and Stanford Jacobs of the Great Central Railway. Lastly, I would like to thank my wife Pat for her patience, encouragement, and not least for checking through the manuscript.

RODGER BRADLEY *Barrow in Furness*

CHAPTER ONE

PIONEER PASSENGER 4-6-0s

The LNER possessed the greatest variety of 4-6-0 locomotive types of each of the four main line railways after 1923, but most of these it had inherited from pre-Grouping companies, designing and constructing only three new types between 1923 and 1947. Among the varied designs of 4-6-0 owned by the LNER were the first passenger types to run on any railway in Britain, belonging to the former North Eastern Railway. The NER, incorporating perhaps the pioneering spirit of the world's first public railway, was a major force in the introduction of the type for passenger service. With Wilson Worsdell as locomotive superintendent, the first 4-6-0 for passenger duties in this country was built for the North Eastern Railway in 1899. Of the other eastern companies, the Great Northern did not build any of this wheel arrangement, while the Great Eastern under Holden saw the arrival of a very successful design. On the Great Central, the greatest variety of 4-6-0 types was to be found, James Robinson producing no fewer than nine different types, all of which passed into LNER ownership.

The pioneer 4-6-0 type in Great Britain was the Jones Goods, which arrived on the Highland Railway in 1894, for freight work. However, while the wheel arrangement had already proved popular in service with a number of overseas railways – notably in the USA, Canada, and Australia – it was not immediately seen in Britain as a contender for express passenger duties. The fashion on a number of railways at that time was for locomotives with large single driving wheels, 4-4-0s, and 4-4-2 Atlantic types. The Atlantic had found favour with the Great Northern Railway, and it had developed an enviable reputation in service, which was perhaps one of the reasons why the company did not progress with the 4-6-0 arrangement for express passenger duties. Further north, in North Eastern Railway territory, there was some need for increasing the power of the passenger locomotives, and the company did initially look in the direction of the 4-6-0 type

for a solution. Despite having pioneered the use of this wheel arrangement in passenger service, the NER later followed the practices of its next-door neighbour, the Great Northern, and adopted the Atlantic type for a principal express passenger design.

On the technical front, there has been the suggestion that even if the 4-6-0 had been introduced earlier than the Jones Goods in 1894, it would have been decidedly less successful, among the reasons put forward for this being the unequal tyre wear of the six-coupled arrangement. In general in the 1890s harder tyres and heavier rails were being introduced by many railway companies, which was obviously beneficial to the operation of the 4-6-0.

Wilson Worsdell had introduced a very successful series of passenger 4-4-0s to the NER, in the M and Q classes of 1892 and 1896. In 1898 an order was placed with Gateshead Works for the construction of 20 new locomotives, ten of which would be the R class 4-4-0, and ten an entirely new design of 4-6-0, the S class. Originally the order was for Q1 class 4-4-0s, but was amended to produce the new design of 4-4-0, and the country's first 4-6-0 for working passenger trains.

The only similarities between the Jones Goods of 1894 and Worsdell's S class were the use of the same size (20 in × 26 in) cylinders, and the wheel arrangement. The coupled wheels of the Highland Railway locomotive were smaller at 5 ft 3 in diameter, with the centre pair flangeless. The boiler dimensions and grate area were only slightly less in the Scottish locomotive than those of the NER design, although the boiler pressure in the S class was substantially higher.

The Highland Railway design was built by Sharp Stewart & Co, and it has been suggested that it could well have been a development of an L class 4-6-0 constructed for India, 11 years earlier in 1883. Unlike the first British goods 4-6-0, none of the pioneer passenger locomotives was rescued for preservation.

Wilson Worsdell's time on the NER also saw the days of the races to the north, in which the east and west coast companies competed with one another for the fastest Anglo-Scottish journey times. Several companies, the NER included, purposely designed locomotives to run at high speeds, but with only light trailing loads, in order to beat times recorded by their rival organisations. By the late 1890s these curious and fascinating events had all but died out, and on the NER it was seen as a time to provide greater power, as passengers demanded comfort rather than speed, requiring increased haulage capacity in the design of passenger locomotives. For this company at least, quickly followed by the Great Western and later by other railways, the ten-wheeled locomotive had arrived.

The North Eastern's new S class 4-6-0 caused quite a stir when it was first unveiled, being considerably larger than any other passenger type then in service. Wilson Worsdell intended that these locomotives would haul trains of between 350 and 375 tons and run 124$\frac{1}{2}$ miles non-stop at average speeds of more than 50mph. With short-travel slide valves, compared with the piston valves used in the new R class 4-4-0s ordered at the same time, the S class had smaller than usual

One of the first British 4-6-0s, North Eastern Railway S class No 760, built in 1906, but in this case surviving only 25 years, to be withdrawn in 1931. *L&GRP*

coupled wheels for an express passenger type of the day. The heavy express passenger trains, which it was intended these locomotives would haul, were for the York to Edinburgh portion of the Anglo-Scottish workings. On this route there were several sections which included gradients severe enough to provide testing runs for the new locomotives and their heavier trains.

Some contemporary press reports recounted successful trial runs with the first member of the class, No 2001, taking 352

NER Class S (LNER Class B13) 4-6-0 – Leading details, as built

Designer	Wilson Worsdell	Grate area	23 sq ft	
Built	6/1899–3/1909			
Withdrawn	10/1928–10/1938	Cylinders		
Running Nos	726, 738–741, 743–763, 766, 768, 775, 1077, 2001–2010	– number	2	
		– dimensions	20 in × 26 in	
Overall length	53 ft 11$\frac{1}{4}$ in	Valve gear	Stephenson, with slide valves (Nos 2001–2008), piston valves for the remainder.	
Overall width	8 ft 10 in			
Overall height	13 ft 1 in			
Wheelbase	26 ft 0$\frac{1}{2}$ in	Tractive effort	24,136 lb	
Coupled wheel diameter	6 ft 1$\frac{1}{4}$ in			
Bogie wheel diameter	3 ft 7$\frac{1}{4}$ in	Axle load	*Tons Cwt Tons Cwt Tons Cwt*	
		– coupled	11 19 19 07 14 19	
Boiler			*Tons Cwt*	
– diameter	4 ft 9 in	– bogie	16 03	
– length	15 ft 0 in			
– tubes, small	204 × 2 in (superheated later)	Adhesive weight	45 tons 5 cwt	
Heating surface				
– boiler	1639.00 sq ft	Weights	*Full*	*Full*
– firebox	130.00 sq ft		*Tons Cwt*	*Tons Cwt*
– Total	1769.00 sq ft	– locomotive	62 08	62 08
		– tender (small)	38 12 (large)	41 02
Working pressure	200 lb/sq in	– Total	101 00	103 10
Firebox		Fuel capacities– coal	5 tons	
– length (o/s)	8 ft 0 in	– water	3,701 gallons (small tender)	
– width (o/s)	3 ft 11 in		3,940 gallons (large tender)	

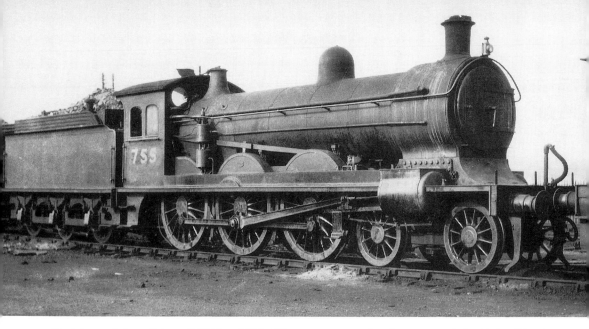

North Eastern Railway S class – LNER Class B13 No 755 – seen from the right-hand side, in early LNER days. Compare this with the other views of S class locomotives – the differences in design of coupled wheel splashers are obvious. *L&GRP*

tons up a gradient of 1 in 78, almost from rest. Starting the same load at the foot of a 1 in 190 gradient, five miles long, the average speed recorded by the summit was 36mph. The hill climbing ability of these locomotives, with their heavyweight trains was by all accounts potentially very good. Yet, some engineers of the day did express doubts on their ability in the speed stakes, with heavy, tightly timed trains. Principal dimensions of the class are given on page 7.

Construction

The foregoing details refer to the locomotive in 'as built' condition, which in many cases altered significantly during subsequent rebuilding and modifications. As early as January 1901 the first locomotive of the class was modified by the rearward extension of the frames, to permit enlarging the original, tiny cab. The length of the locomotive, which imposed such severe limits on the size of the cab, only affected the first three.

In the new express passenger 4-6-0 the boiler, large by contemporary standards, was built up from three rings 4 ft 9 in diameter, and 15 ft 0 in long, from $^9/_{16}$ in thick steel plates. The diameter over the outside of the cleading was 5 ft $0^1/_2$ in. In the boiler 204 2 in diameter tubes were installed; the length be-

tween tubeplates was 15 ft $4^1/_8$ in. The last two locomotives of the first batch of class S were built with saturated boilers, but piston valves instead of slide valves were preferred. After 1906 only 225 $1^3/_4$ in small tubes were fitted, reducing the heating surface to 1,578 sq ft.

A conventional copper inner firebox was fitted, 7 ft $3^1/_2$ in long, and stayed to the steel outer firebox plates with copper stays. The inner firebox was 5 ft 8 in deep at the front, 4 ft 8 in at the back, and 3 ft $2^1/_2$ in wide at the bottom. The firegrate with an area of 23 sq ft sloped downwards at a shallow angle from just in front of the trailing coupled axle. Outside, the firebox was 8 ft 0 in long, with a round top, surmounted by Ramsbottom type safety valves, with the sides of the box waisted-in to fit between the locomotive frames.

At the opposite end of the locomotive a short smokebox only 3 ft $3^1/_2$ in long was fitted. Superheating was not then provided, so the extra space was not required in the smokebox to accommodate superheater headers, at least, not in the original design. A curious feature of these locomotives was the adoption of the same size blast nozzle, as on the R class 4-4-0s, only 5 in diameter, which doubtless contributed in some way to the less than satisfactory performance of these new engines. Superheaters fitted to Nos 2001 to 2010 at a later date required a longer smokebox, whose forward extension covered a hinged plate which previously gave access to the cylinder valve covers at the leading end of the locomotive. Superheating was the only major change to occur in the boilers of the B13 class.

If a boiler required to be taken off a locomotive for repair or overhaul, the NER practice was to mount a spare in its place. As locomotives came in for overhaul, their boilers would be interchanged. This practice was maintained as successive batches of class S or B13 were introduced between 1906 and 1909, and six more spare boilers were constructed. The availability of these spare boilers allowed more than one locomotive to be overhauled at a time, and the replacement of some of the earlier boilers.

In 1913 superheating was introduced to the class, with the construction of seven more boilers, permitting the scrapping of more of the older ones. Darlington, where more of this design of boiler were built in the 1920s, produced the first superheated boilers. Schmidt type superheaters were installed, reducing the heating surface of the small tubes to 1265.6 sq ft, with the superheater itself providing 378.0 sq ft of additional heating surface. The eighteen $1^3/_{32}$ in outside diameter elements, were contained in $5^1/_4$ in diameter flue tubes. Boiler working pressure in the superheated locomotive was reduced to 160 lb/sq in from the 175 lb/sq in carried by the majority of the class. Only the first three locomotives, Nos 2001–2003 had boilers pressed to work at 200 lb/sq in, which gave these a nominal tractive effort of 24,136 lb, coming down to 21,119 lb and 19,309 lb in the later variations.

Frames, Wheels and Motion

It was a requirement that the North Eastern Railway's first passenger 4-6-0 could be turned on existing 50 ft 0in diameter turntables; with a short tender, the total wheelbase was kept down to 48 ft $4^3/_8$ in. While certain dimensions of the locomotive – the wheelbase for instance – were fixed by the diameter of the coupled wheels, the overall wheelbase of locomotive and tender was kept within these restricting dimensions by taking some 2 ft 0 in off the rear of the cab and footplate. The tender, too, was shorter than might normally have been expected, although by only 18 in. The most obvious feature in this shortening of the design was to be seen in the diminutive cab, with its single side window.

The locomotive main frames themselves were in these first examples only 33 ft $4^1/_4$ in from drag beam to front buffer beam. Mild steel plate 1 in thick was the material used, with inside bearing axleboxes only, and stayed apart at 4 ft 0 in between the inside faces of the frames. From just in front of the leading coupled wheels they were narrowed in slightly under the smokebox, and over the bogie to the front buffer beam. The main axleboxes, 8 in diameter, and 9 in long were supported by coil springs on the leading and driving coupled axles, and underhung leaf

Unrebuilt B12 No 8552 on former Great Eastern metals, still sporting its decorative valance, but not the disfiguring encumbrance of the feed water heating gear. *L&GRP*

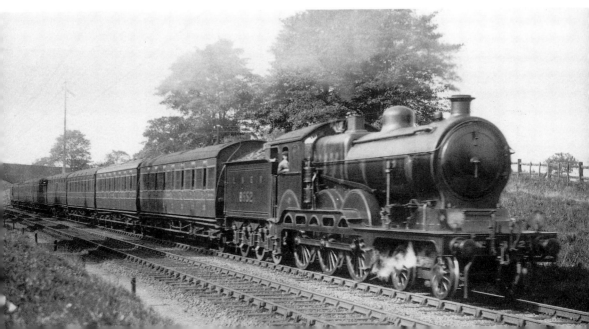

springs on the trailing axle. The frame plates were curved over the axle positions, and at the leading end over the bogie pivot centre and cylinder positions.

Coupled wheel diameter was 6 ft 1¼ in, with 16 spokes and 5½ in wide tyres. The centre pair of wheels on a number of the first S class locomotives were originally flange-less, although this was subsequently found to be unsatisfactory in service, and normal flanged driving wheels were fitted. All tyres on coupled and bogie wheels were 3 in thick. The coupled wheels were fitted on to 7¾ in diameter axles. The leading bogie was fitted with inside bearings, with 12-spoke 3 ft 7¼ in diameter wheels on a wheelbase of 6 ft 6 in. Leaf springs provided suspension and side control, with an equalising beam between the two axles.

In the original locomotives the 20 in by 26 in cylinders were operated by Stephenson valve gear, with short-travel, balanced slide valves. The cylinders were carried outside the frames, with the steamchests, valves, and motion inside. The connecting rods were deeply fluted, and some 10 ft 6 in in length, with double slidebars and crosshead. The pistons were fitted with long tail rods, and a head carrying two rings. The first eight loco-motives were built with slide valves having a maximum travel of 4²¹/₃₂ in, 1¹/₈ in lap, and a lead of ¹/₈ in. The last locomotive in the first order, Nos 2009/2010, were built with piston valves, and all subsequent orders for S class 4-6-0s were similarly fitted. The earlier slide valve locomotives received new cylinders and slide valves over a number of years – No 2001 in January 1901, and the last to be converted, No 2002, in July 1924.

General Development

The distinctive feature of the new S class was the extremely short cab, with its single side window. North Eastern Railway locomotives had secured a reputation for some of the most commodious cab designs on all Britain's rail-ways, yet this class apparently bore in mind the troubles of 1884 with the McDonnell 4-4-0s, and their excessive length. The S class cabs were initially shorn of 2 ft 0 in of their desirable length. The distance from the trailing coupled axle to the rear buffer beam was only 4 ft 10 in, of which barely half was available for the footplatemen to stand on, and resulted in some peculiarly cramped

working conditions. The screw reverse fitted would doubtless have been easier to operate from the tender, not the locomotive, since it reached back almost beyond the cab side sheets! This considerable disadvantage was discovered almost as soon as the locomotives were built; only the first three were affected, and those were rebuilt into normal style in January, June, and September 1901. Two side windows had been proposed for even the shortened cabs, but this idea was dropped in favour of the single window. Another curi-ous effect in these short locomotives was the adoption of a type of drawgear between loco-motive and tender that resulted in an in-crease in the locomotive-to-tender gap of 4 in by comparison with the usual arrangement.

The early examples were fitted with brass-rimmed chimneys, and large Ramsbottom type safety valves on the firebox casing. Later members of the class carried a chimney with a less pronounced taper but with *ca-puchon* (smoke deflector) on the front rim, which were referred to in the North East as 'windjabbers.'

On the right-hand side of the locomotive the Westinghouse air pumps for the braking system were fitted just in front of the cab, between the driving and trailing coupled wheel splashers. The majority were fitted with dual braking systems – Westinghouse air brakes for the locomotive and former NER trains, and vacuum brake equipment for working vacuum-fitted rolling stock. The last ten locomotives built to this design in 1908/1909 had Westinghouse equipment only from new, but were later modified to be dual-fitted in common with the rest of the class. All had their front end connections fitted below the front buffer beam, but sub-sequent alterations saw the installation of the traditional swan neck vacuum pipe fittings. Brake shoes were carried ahead of the treads of each wheel, on single hangers. Gravity sanding was provided to the front of the lead-ing coupled wheels, and behind the trailing pair. When a further 20 locomotives were built in 1908/1909 steam sanding was adopted for the front wheels, although this reverted in later days to gravity feed, with the exception of two or three locomotives which retained steam sanding equipment.

The locomotive running boards were almost continuous from cab to front buffer beam, broken only by the intrusion of the cylinders, and with a narrow hanging bar or valance. On the first locomotive the coupled wheel splashers were very wide, extending to within an inch or two of the footplate edge. Later members of the class had wheel splashers similar to those fitted on the Atlantic types, which were much narrower, and the rear splasher was not merged into the cab side sheets.

A less obvious detail difference, which was seen only on No 2003, was the inclusion from new of the Younghusband Patent Valve Gear. This fitting was in the nature of an experiment, following trials of the gear on other North Eastern Railway locomotive types, including a 2-4-0 and a number of 0-4-4 tank engine designs. In arrangement, it was a modified form of the stationary link motion developed by J. V. Gooch, with the intention of providing the maximum possible port opening as early as possible. This valve gear was heavier than the usual Stephenson link motion, and the reach rod was altered to give increased leverage on the weigh bar. On No 2003, this gear could be detected by the cranked reach rod, compared with the standard straight version in all other members of the class. Similarly, changes were made in the cylinder lubrication system, where originally two pipes from small lubricators attached to the inside faces of the mainframes, in front and behind the cylinder castings, were replaced by a mechanical lubrication arrangement.

The smokebox door was closed by a wheel and handle, although a two-handle arrangement was used on some locomotives, which later reverted to the handwheel fastening, only to be replaced again later by a two-handle fastening. While it has been suggested that these locomotives originally had their smokebox doors fastened by the two-handle arrangement, contemporary drawings of the type, including arrangement diagrams, show the wheel-and-handle fastening.

Construction

Despite not meeting the standards of the duties for which they were originally intended, 40 S class (later LNER B13 class) locomotives were built at the North Eastern Railway Gateshead Works between 1899 and 1909. All were built under Wilson Worsdell's authority, and it must have been very

satisfying to see No 2006 win the *Grand Prix* and a gold medal at the Paris Exhibition of 1900.

NER Class S (LNER Class B13)
Construction and withdrawal

Running No	Built	Superheated	Withdrawn
2001	6/1899	2/1916	6/1931
2002	6/1899	7/1924	7/1931
2003	9/1899	5/1920	7/1931
2004	12/1899	4/1921	8/1928
2005	12/1899	11/1916	11/1928
2006	12/1899	4/1918	6/1931
2007	3/1900	6/1916	10/1928
2008	5/1900	4/1917	12/1929
2009	6/1900	7/1915	7/1931
2010	6/1900	4/1916	7/1931
726	4/1906	11/1915	12/1936
740	4/1906	7/1914	9/1932
757	4/1906	3/1918	5/1932
760	5/1906	4/1915	3/1931
761	6/1906	11/1924	9/1934
763	6/1906	1/1916	4/1929
766	6/1906	12/1916	10/1931
768	6/1906	8/1918	5/1929
775	8/1906	12/1920	8/1936
1077	8/1906	6/1918	11/1931
738	6/1908	8/1916	7/1938
739	6/1908	2/1925	7/1932
741	6/1908	5/1917	1/1930
743	7/1908	10/1917	5/1932
744	7/1908	11/1915	12/1931
745	8/1908	12/1921	12/1931
746	8/1908	3/1918	11/1931
747	9/1908	10/1920	8/1932
748	9/1908	3/1915	10/1938
749	10/1908	11/1915	4/1930
750	11/1908	1/1915	11/1932
751	11/1908	11/1913	5/1936
752	11/1908	3/1920	6/1934
753	12/1908	11/1916	10/1938
754	1/1909	7/1922	12/1936
755	1/1909	8/1917	2/1934
756	1/1909	11/1919	9/1934
758	2/1909	11/1915	5/1930
759	3/1909	9/1923	10/1938
762	3/1909	6/1917	5/1937

Notes:
1 All locomotives were built at Gateshead Works
2 No 761 was transferred to service stock in September 1934 until May 1951 for use as a counter-pressure locomotive, without superheater. It was renumbered 1699 in October 1946.
3 Allocation in 1923 was:

Hull (Dairycoates)	15
Tweedmouth	8
Heaton	8
Leeds (Neville Hill)	6
Blaydon	3

All the locomotives were constructed at Gateshead Works. Darlington Works built 37 superheated boilers for the class between 1913 and 1927. Most major dimensions of the various batches remained relatively unaltered throughout the building period, with the exception of the first three, short-wheel-base locomotives. By the time the LNER had taken over the eastern railways following the Grouping of 1923, detail alterations had changed some of these standards – the wheel-base, the boiler working pressure, and trac-tive effort.

Seven of the third batch and eight of the final batch, had boilers fitted with steel tubes, as did the replacement boilers built at Darlington. All 20 of the first two batches had copper tubes, while the third batch was fitted with patent Hornish boiler cleaning equipment. The introduction of Schmidt pattern superheaters, with the larger smoke-boxes, necessitated changes to the valve in-spection covers at the front end. In the superheated locomotives access was gained by hinges on the plating between the front frames, where previously two small covers were provided just in front of the smokebox front plate.

Tenders
The design of tender fitted to the first batch of class S locomotives was very short, and although it planned to equip all the loco-motives with this tender, not all may have received one. No 2006, the gold medal loco-motive, was exhibited in Paris with a normal style of tender. The wheelbase of these short tenders was 12 ft 0 in, equally divided, with 3 ft 9¼ in diameter wheels, double frames, outside bearing axleboxes, overhung leaf springs, and water pick-up gear. Although all the tenders paired with the S class carried five tons of coal, the short tenders had 8 in shorn off the length of the water tank and carried only 3,701 gallons of water. Initially, the tender sides were topped by two coal rails, running around three sides. Later, four rails were common, reduced to three beyond the rear of the coal space and around the tender rear, and plated over where the rails surrounded the coal bunker. These changes were evident from 1909.

The basic design of the larger tender, fit-ted to most of the class was essentially the same. The wheelbase of the larger tender was 12 ft 8 in, again equally divided, with the tank capacity raised to 3,940 gallons, while the weight in full working order was 41 tons 2 cwt compared with the 38 tons 12 cwt of the short tender. In both cases toolboxes were provided on either side of the coal space, with the water tank vents at the forward end of the tender.

Operations
Although these large and potentially power-ful 4-6-0s, the progenitors of many hundreds of others, were intended for the NER's heav-iest express passenger work, they were not as successful as their contemporaries, the R class 4-4-0s. Troubles with the running gear on the new design have been associated with inadequate side clearances in the coupled wheel bearings and rods, while on an early trial run, an overheating eccentric prevented the locomotive from showing its paces to the full. With a trailing load of 350 tons on this trial, held at Flying Scotsman timings, no problems were encountered with the boiler's steaming ability or capacity. The old NER S class boilers steamed well under heavy work-ing conditions, in spite of the disadvantage perhaps, that with a barrel almost 4 ft 0 in longer, the blast nozzle diameter of 5 in was the same as that used on the much smaller R class 4-4-0s. The company's plans to elim-inate double heading on the Newcastle to Edinburgh section did not come to fruition, with rather disappointing results from fur-ther trial runs. For express passenger duties, the R class 4-4-0, and class V Atlantics of 1903 were preferred, even though a new design of 4-6-0 was introduced (the S1 class), only two years after this pioneer locomotive. The class S, including the later batches, were employed on excursion and perishable goods workings, on both the NER and later, as Class B13 under LNER ownership, through-out almost all their working lives.

With the exception of Nos 2002, 761, 739 and 759, these locomotives were superheated by the North Eastern Railway. All were equipped with Sir Vincent Raven's fog signalling apparatus. Following the Group-ing, a number received the Ross pop safety valves, seated on a truncated version of the original Ramsbottom safety valve cover. Minor alterations during the LNER period included changes to the front end vacuum brake connections, while Nos 740, 755, and 775 were fitted with LNER standard buffers. The standardised dimensions issued for these locomotives by the LNER differed in a number of aspects from those of the original, listed earlier, and for the B13 class of the LNER are noted in the accompanying table.

LNER Class B13 4-6-0 – Later standard details

Wheel arrangement	4-6-0
Wheelbase (locomotive only)	26 ft 0½ in
Bogie wheel diameter	3 ft 7¼ in
Coupled wheel diameter	6 ft 1¼ in
Tender wheelbase	12 ft 8 in
Tender wheel diameter	3 ft 9¼ in
Length overall	61 ft 0¾ in
Cylinders	
– number	2
– dimensions	20 in × 26 in with 8¾ in diameter piston valves
Boiler	
– diameter	4 ft 9 in LNER Diagram No 54
– length	15 ft 0 in
– tubes, small	126 × 1¾ in o/d
– tubes, large	18 × 5¼ in o/d
Superheater elements	18 × 1³/₃₂ in o/d
Heating surface	
– tubes	1263.00 sq ft
– firebox	120.00 sq ft
– Total	1383.00 sq ft
– superheater	276 sq ft
Working pressure	160 lb/sq in
Firebox length	8 ft 0 in
Grate area	23.0 sq ft
Tractive effort	19,309 lb
Fuel capacities– coal	5 tons
– water	3,940 gallons
Weights in working order	
– locomotive	64 tons 06 cwt
– tender	43 tons 10 cwt
– Total	107 tons 16 cwt
Adhesive weight	48 tons 02 cwt
Maximum axle load	19 tons 14 cwt

The Great Central Railway war memorial locomotive as LNER No 6165 *Valour* at Nottingham Victoria in 1931 – an attractive and powerful looking design, whose appearance was not borne out in service with the LNER as the B3 class. *L&GRP*

Two locomotives were used by the chief mechanical engineer's department as counter-pressure locomotives. The first, for only a year in 1934 was No 756, still in running stock, while in 1935 No 761 was transferred to service stock for this work, where it remained until withdrawal in May 1951 as No 1699.

Although intended to be a passenger type, the S class locomotives were used throughout their careers for mixed-traffic work. In LNER days their duties involved parcels, perishable goods, stopping passenger, occasional goods or mineral, and excursion workings. Most of the services which were in the charge of B13s could be seen in the North East, on the main lines between Doncaster and Edinburgh, and Leeds to Hull and Scarborough, while a number had regular cross-country jaunts to the west coast and Carlisle. Although the B13s never achieved such auspicious duties as Royal Train working again – which Nos 2009 and 2010 achieved in 1900 – they were still in the 1930s to be found on occasions at the head of such trains as the Sunday Pullman, from Leeds to Harrogate. Withdrawals began in 1928, with No 2004 the first to go in August, followed by No 2007 in September the same year. None of the class remained in running stock by the outbreak of World War II; in fact, by 1934, only nine still survived. With the exception of the counter-pressure locomotive No 761, the majority of the locomotives were withdrawn in the years 1930–1932, with 1931 being a particularly bad year for the B13. The last to go were Nos 748, 753 and 759, with the last having a particularly ignominious end when it failed completely on taking-over a Hull to Barnsley fast fish train just out of Doncaster, and was dumped in a nearby siding with the fire dropped! Only the replicas of the gold medals carried in service by No 2006 were salvaged from the pioneer 4-6-0 design; those were handed over to the York Railway

Museum for preservation in 1931. The class S 4-6-0 was not entirely without saving qualities, since more passenger locomotives of this type were being built by the North Eastern Railway and other contemporary railway companies, with some of the designs of the pre-Grouping companies inherited by the LNER perpetuated for many years.

Contemporary Developments and Successors
On the North Eastern Railway three more classes of 4-6-0 were built under Wilson Worsdell's and Sir Vincent Raven's authority, with the last design, the S3 (later LNER B16) built as LNER locomotives in the first few years after the 1923 Grouping. The first of the NER's second generation 4-6-0s, class S1 (LNER B14) were built for passenger work, notwithstanding the poor record of the first Worsdell design, and were equipped with 6 ft 8 in coupled wheels. Raven's contribution to the new range of 4-6-0s included two mixed-traffic types, although the final design was originally hailed as a fast goods class.

Between 1899 and 1924 the NER and LNER built no fewer than 172 locomotives of 4-6-0 wheel arrangement, in four classes, B13, B14, B15, and B16. Other companies were not idle in the building of the new

designs, in particular the Great Central Railway under J. G. Robinson's supervision, where no fewer than nine different designs emerged between 1902 and 1922. Further east, the Great Eastern under Holden contributed only one design during this period, which saw a variety of successful rebuilds, particularly under Gresley's direction in LNER days. Surprisingly perhaps, while Gresley was cme of the LNER, only one new design of 4-6-0 was built, the B17 Sandringham class, intended as a replacement for the B12s. There was some financial restriction on the LNER, with a locomotive policy that did not go in for the scrap-and-build practices of other railways – more modest rebuilding and alteration of existing, sound 4-6-0s designs was pursued.

In total, no fewer than 17 different designs of 4-6-0 were in LNER service between 1923 and 1947, with the majority built by the pre-Grouping companies. Most came from the Great Central Railway under the direction of J. G. Robinson, where four new classes were outshopped before the pioneering North Eastern Railway had completed its first two designs. The first of these GCR designs became LNER class B5, of which 14 were built between 1902 and 1904, and in 1903 the two members of class B1 (later reclassified

Summary of LNER 4-6-0 classes

LNER Class	Pre-Grouping railway	Pre-Grouping class	Quantity built	Building period
B1 (Later B18)	GCR	8C	2	1903
B2 (Later B19)	GCR	1	6	1912–1913
B3	GCR	9P	6	1917–1920
B4	GCR	8F	10	1906
B5	GCR	8	14	1902–1904
B6	GCR	8N	3	1918–1921
B7	GCR	9Q	38	1921–1922
B8	GCR	1A	11	1913–1915
B9	GCR	8G	10	1906
B12	GER	S69	81	1911–1920/1928
B13	NER	S	40	1899–1909
B14	NER	S1	5	1901
B15	NER	S2	20	1911–1913
B16	NER	S3	70	1919–1924
B17	—	—	73	1928–1937
B1*	—	—	410	1942–1952
B2**	—	—	10	1945–1949

Notes:
* Originally these were to be classified as B10, but Thompson later issued instructions that they were to be B1, and the ex-GCR 4-6-0s reclassified B18.
** These were rebuilds of the B17 Sandringhams, with B1 type boilers; the ex-GCR locomotives were reclassified B19.

B18) appeared. There then followed a spate of 4-6-0 building by the GCR, with another seven types outshopped by the company before the end of 1922.

The most numerous of the Great Central designs were the 38 four-cylinder class 9Q (LNER class B7) constructed by Gorton Works, the Vulcan Foundry, and Beyer Peacock in Manchester, between 1921 and 1922. Only 28 were actually built by the GCR, with the rest appearing as LNER locomotives in 1923–1924. This last of the many Robinson 4-6-0s, although renowned for its heavy coal consumption, was possibly one of the better designs, and in common with seven other GCR types, survived to become British Railways stock.

A major force among the east companies, the Great Northern Railway was not following the fashion of the early 1900s and building 4-6-0s. H. N. Gresley, the chief mechanical engineer, offered no such type to the LNER at the Grouping in 1923. As the most well-known of the LNER chief mechanical engineers, Gresley will be remembered for his illustrious Pacific types, although his development of 4-6-0s as suited the LNER locomotive policy provided some interesting changes.

Gresley presided over the introduction of one entirely new 4-6-0 type, the B17 Sandringham class, and the rebuilding of many of the former NER, GCR, and GER 4-6-0s.

The last-mentioned company's contribution to LNER locomotive stock in this category, consisted of the solitary S69 class locomotive, which were built under S. D. Holden's authority over a ten-year period between 1911 and 1921. In total, 81 were constructed: the LNER inherited 71 from the GER and ordered ten more from Beyer Peacock in 1928. The B12s under LNER management underwent a number of rebuilds and modifications, but survived in such numbers that 69 were included in British Railways stock in 1948, with the last member of the class going to the scrapyard in 1960.

While the 4-6-0 type may not seem to have reached the peaks of popularity, at least initially, on the LNER as it had with contemporary companies, the pioneering efforts of the NER and others resulted in substantial numbers and designs remaining in service for many years.

GCR Class 1A No 279 *Earl Kitchener of Khartoum*, later to become LNER Class B8 in GCR livery at Leicester. Note the Robinson top feed on the front boiler ring. *L&GRP*

Gresley's only essay in 4-6-0 locomotive design, albeit with the considerable assistance of the North British Locomotive Company, was the B17 or Sandringham Class. Here seen in close-up, is the left side of No 2817 *Ford Castle. Gresley Society*

The culmination perhaps of LNER 4-6-0 design and experience was Thompson's B1 design. Here No 61218 seen at Darlington in 1953, in early BR lined black livery, looks a most attractive steam locomotive. *L. R. Peters/Gresley Society*

CHAPTER TWO

FURTHER DEVELOPMENTS ON THE NORTH EASTERN

Despite the relatively disappointing performance of the first passenger 4-6-0s, the S class, overshadowed in everyday service by the highly successful R class 4-4-0, another design of 4-6-0 was introduced in 1900. The new wheel arrangement, it was accepted, could be suitable for passenger duties, and barely a year after the first British passenger 4-6-0 was born, Wilson Worsdell brought into being the S1 class, later to become LNER B14.

By the time of the 1923 Grouping the North Eastern had built and was operating four new classes of 4-6-0. Classes S, S1 and S2 had two outside cylinders. The largest class, S3 – 38 locomotives, with a further 32 built by the LNER after 1923 – had three cylinders.

Unrebuilt B16 No 61416 on a common turn of duty for the class in BR days. *G. W. Sharpe*

As LNER locomotives, the two follow-up 4-6-0 classes produced under Worsdell and the final design of Vincent Raven became classes B14, B15 and B16. Not all of these survived the LNER years, with the B14s all going before World War II, and the B15s mainly during the 1944–1946 period. The 4-6-0 design produced in Raven's days, class S3 in NER conventions, survived much longer. All except one of the 38 built by the NER, and the 32 produced in LNER days became British Railways stock, almost seeing the end of steam traction on the Eastern Region.

NORTH EASTERN RAILWAY
S1 CLASS 4-6-0 – LNER CLASS B14

Persisting with the idea of the 4-6-0 wheel arrangement, Wilson Worsdell's next foray into the design of this type of passenger loco-

motive, for the heaviest work, resulted in a type that was even larger than its predecessors. The contemporary technical press, in particular *The Engineer*, was very impressed by their size. In the 26 July 1901 issue, Rous-Marten described them as 'huge', and some nine tons heavier than Aspinall's 'gigantic' 1400 class! On the same page, the bottom right-hand corner carried an invitation to tender from the Government of New South Wales for the building of the Sydney Harbour Bridge, whose load-carrying capacity was tested with many steam locomotives. As one giant steam locomotive of the day was just appearing, so a giant in a different field, and one that is still with us, was about to take shape. It was not however an omen of good fortune for the NER's S1 class 4-6-0, which numbered only a handful of locomotives, and were quickly relegated to duties for which they were not intended.

In presenting this huge new steam locomotive to the world, one of Worsdell's main alterations to the previous design was to raise the centre line of the boiler to 8 ft 6 in above rail level and to increase the diameter of the coupled wheels by 7 in. In original form with these alterations, they were largely similar to the pioneer passenger S class 4-6-0.

Boiler design

The S1 class 4-6-0s were built with saturated boilers, in two rings, from $^9/_{16}$ in steel plate, 15 ft 10½ in long, and 4 ft 9 in diameter. Initially, the boilers carried 200 lb/sq in pressure, but this was soon reduced to 175 lb/sq in which was maintained as standard, even after superheaters began to be fitted from 1913 onwards. Originally, 196 2 in diameter tubes were installed, but from the date of superheating this was reduced to 126 $1^3/_4$ in diameter tubes, and eighteen $5^1/_4$ in diameter superheater flues, each carrying a single $1^3/_{32}$ in diameter element.

Six saturated boilers, including a spare, were built for the new 4-6-0s. However, only five were replaced with boilers carrying the new Schmidt superheaters between 1913 and 1917, and consequently no exchange of boilers was readily possible utilising the spare boiler after that time. The reduction in boiler pressure from 200 lb/sq in to 175 lb/sq in, affected the locomotives' tractive effort, bringing the figure down from 22,069 lb to the 19,310 lb quoted in the table of leading

NER Class S1 (LNER Class B14) 4-6-0 – Leading details, as built

Designer	W. Worsdell
Built	12/1900–8/1901
Withdrawn	6/1929–4/1931
NER running numbers	2111–2115
LNER running numbers	2111–2115

Wheel diameter	
– coupled	6 ft 8¼ in
– bogie	3 ft 7¼ in

Wheelbase	27 ft 6 in

Axle load	Tons Cwt	Tons Cwt	Tons Cwt
– coupled	15 07	19 10	17 02
	Tons Cwt		
– bogie	15 03		

Boiler	
– diameter	4 ft 9 in
– length	15 ft 10½ in
– tubes small	193 × 2 in o/d

Cylinders	
– number	2
– dimensions	20 in × 26 in

Heating surface	
– tubes	1639.00 sq ft
– firebox	130.00 sq ft
– Total	1769.00 sq ft

Working pressure	175 lb/sq in

Firebox	
– length o/s	8 ft 0 in
– width	3 ft 11 in

Grate area	23.00 sq ft

Tractive effort	19,310 lb

Fuel capacities	
– coal	5 tons
– water	3,940 gallons

Weights	Full	
	Tons	Cwt
locomotive	67	02
tender	41	02
Total	108	04

details. There was also some slight difference in heating surfaces, with the different methods of calculation used by the North Eastern and LNER companies. The NER for instance quoted a figure of 294 sq ft for the superheater heating surface, whereas the standard LNER methods gave 390.1 sq ft on official diagrams and documents.

The outer firebox casing, of steel, was 8 ft 0 in long, with a maximum width of 4 ft 9 in, built up from $^5/_8$ in thick plates, with its round top flush with the boiler barrel. The

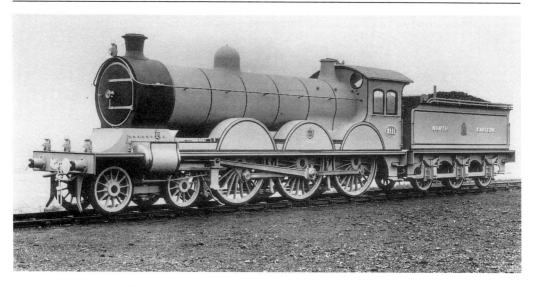

Following on from the first NER essay into the design of a 4-6-0 type, was the larger S1 class, later to become Class B15 in LNER days. Like all NER locomotives, the cab was a far more comfortable affair than that provided by the majority of the company's contemporaries. *National Railway Museum*

copper inner box was 7 ft 3^1/$_2$ in long, 3 ft 2^1/$_2$ in wide, with a grate area of 23 sq ft. The inner firebox was stayed to the outer casing using 1^1/$_8$ in diameter copper stays at 4 in centres. At the other end of the locomotive a simple short smokebox housed only the 5 in diameter blastpipe. The smokebox was later lengthened to accommodate the superheater header. Originally, the smokebox door was fastened by means of a wheel and handle, but in later years Nos 2111, 2114 and 2115 were fitted with the more commonplace two-handle arrangement.

Cylinders and Motion
The Worsdell S1 class 4-6-0s had a number of similarities with the earlier design, although there were also many dimensional changes. Some of these changes were made to accommodate the much larger coupled wheels. The locomotive frames were 1 in thick mild steel plate. The coupled axles were positioned at 7 ft 7 in apart, and the centre line of the bogie was 9 ft 1 in from the leading axle, giving a total wheelbase of 27 ft 6 in.

Main axleboxes were in steel, 9 in diameter by 9 in long, with 6 in by 9 in journals for the bogie wheels, while the wheels were 6 ft 8^1/$_4$ in coupled, and 3 ft 7^1/$_4$ in diameter bogie.

Locomotive main suspension again showed some variety in the new 4-6-0, with under-hung leaf springs on the trailing axle only, and helical springs for the driving and leading coupled axles. Like the pioneer design, brake shoes were carried ahead of the wheels, with sanding applied in front of the leading pair, and behind the trailing coupled wheels. As standard practice on the NER, Westinghouse equipment was installed for locomotive and train, although a vacuum ejector and train pipe were fitted for alternative use if required.

The two 20 in by 26 in stroke cylinders were carried outside the frames, driving onto the centre pair of coupled wheels, as previously. Inside the frames, though, the earlier slide valves had given way to 8^3/$_4$ in diameter piston valves with a maximum travel of 4^{21}/$_{32}$ in, operated by Stephenson valve gear. The coupling and connecting rods were deeply fluted, and the latter were some 11 ft 4^1/$_2$ in long, and fitted with a marine type big-end. Other similarities with the predecessor class S 4-6-0 were the degree of inclination of the cylinders, and the provision of tail rods to the pistons.

General Details
The new 4-6-0s were not foreshortened as their predecessors had been, and from new they included a commodious cab with two side windows. At running board level what later became LNER class B14 were similar to the B13, although much larger coupled

wheel splashers were fitted, and minor alterations were made to the cab and front footsteps. The boiler was surmounted by the steam dome, housing the regulator, with Ramsbottom type safety valves on the firebox roof. Nos 2111/13/14 carried these fittings throughout their working lives, but Nos 2112 and 2115 from the same batch were fitted with modern pop type valves by the LNER. The chimney was a tapered cast-iron design, later fitted with a 2 in high smoke deflector, or 'windjabber' as it was known on the North Eastern Railway. Another fitting which became standard was the Raven fog signalling apparatus, with the striking gear carried on these locomotives just in front of the driving axle, between the frames.

Tenders paired with the new locomotives were larger, but carried on a six-wheeled underframe, although there is some disparity in dimensions and capacities quoted for this design. As built, the S1 class was reported as having a tender weighing 40 tons exactly in working order, carrying 4,729 gallons of water and 5 tons of coal. LNER figures show a tender weighing 41 tons 2 cwt, and carrying only 3,940 gallons of water. Initially, two open coal rails were fitted, although this was subsequently increased to four, and plated over around the coal space.

Construction and Operation
All five locomotives were built at the North Eastern Railway Gateshead Works, with No 2111 coming out in December 1900, and the remainder in June and August the following year. According to contemporary press reports, the new locomotives introduced by Wilson Worsdell, arrived on the back of the sound work already accomplished by the earlier design, and to determine whether larger wheels would be an advantage. The S1 (LNER B14) class was once again intended for the heaviest Anglo-Scottish express passenger work, to eliminate double-heading on the York to Edinburgh section. The following table shows the building and withdrawal dates:

Running No	Built	Superheated	Withdrawn
2111	12/1900	10/1913	7/1929
2112	6/1901	4/1916	4/1931
2113	6/1901	6/1915	10/1930
2114	6/1901	5/1915	10/1930
2115	8/1901	4/1917	6/1929

In service, they were originally allocated to the heaviest passenger trains between York, Newcastle and Scotland, but as early as 1907 they were replaced by the famous Atlantic types and the highly successful R and R1 class 4-4-0s. By the time World War I arrived, the S1s were already relegated to main line goods work, in particular the Anglo-Scottish fish trains.

At the time of the Grouping in 1923, they were based at Gateshead, and in that year only No 2115 was reportedly still at work on passenger duties. In 1924, the class was reallocated to Hull (Botanic Gardens) and subsequently Dairycoates, where the majority ended their lives on freight, including pick-up goods, and the occasional extra passenger workings. Their withdrawal came about during the years of economic depression in the 1930s, and was settled as the boilers reached the limit of their lives, and became due for renewal.

NER CLASS S2
(LNER CLASS B15) 4-6-0

Vincent Raven succeeded Wilson Worsdell in the post of chief mechanical engineer on the NER in 1910, and under his authority a third new 4-6-0 design appeared, but with many differences from the earlier types. Twenty locomotives of this first mixed traffic class were built at Darlington Works between 1911 and 1913. A change in policy, with the most powerful types adopting three-cylinder propulsion, may have been in some way responsible for the introduction of only 20 locomotives. The order for ten locomotives similar to class S was placed in May 1911, and the use of larger boilers on other NER types was successful enough to warrant similar improvements to the S type design. Accordingly a revised order with other minor alterations in the details of the design produced ten 4-6-0 locomotives, with a repeat order for ten more in February 1912.

Again, in original form, no superheaters were fitted to the first seven locomotives, until company policy decided on the extent of superheating for its motive power, while locomotives of this class from No 797 onwards had superheaters installed from new. A major departure from conventional locomotive design was the fitting of Stumpf Uniflow cylinders to No 825, the last S2 (LNER B15) to be built. As with a number of experiments which sought to improve the efficiency and economy of the steam locomotive, this particular idea, although achieving some popularity in Europe, did not show any significant advantage for the NER.

While the new locomotives had much larger boilers than their predecessors, very few changes were made to the mechanical design, which came out on the whole much the same as S class, LNER B13 4-6-0. Ironically, the new mixed-traffic locomotive suffered similar steaming problems, and were not particularly popular with enginemen, who were unfamiliar with their characteristics. Development of the 4-6-0 type on the eastern railways was progressing, but the popularity of the Atlantic type had perhaps slowed that development, along with early teething troubles with the six-coupled designs. Some of those early difficulties were not always attributable to the new wheel arrangement and mechanical details. Although 20 locomotives were built between 1911 and 1913, they were subsequently replaced by the three-cylinder class S3, LNER B16 type, introduced in 1919.

Boiler Design

The new larger boilers fitted to the B15s were constructed in three rings, and measured 5 ft 6 in diameter, and 15 ft 0 in between tube-plates. In LNER days the official dimension for this latter, 15 ft $7^1/_2$ in, included an extension of $7^1/_2$ in ahead of the smokebox tubeplate, to reach the saddle. The first batch of locomotives with saturated boilers had 268 2 in diameter tubes, providing a heating surface of 2153 sq ft. As the superheater fitted locomotives appeared, with a form of header and installation of elements developed by J. G. Robinson on the Great Central Railway, the number of small tubes was reduced to 146 at 2 in diameter. The twenty-four $1^3/_{32}$ in diameter superheater elements provided 545 sq ft of heating surface as built, although later changes by the LNER, reduced this to either 361 sq ft or 402 sq ft. A further reduction in the number of small tubes on the superheated locomotives, to no more than 90, was made with the intention of increasing the degree of superheat by exposing the large tubes to more heat from the firebox gases as they passed through the boiler. By 1923, the LNER's inherited B15 4-6-0s had a total evaporative heating surface of only 1369 sq ft, with the small tubes providing a mere 723 sq ft. From 1913, superheated locomotives fitted with Robinson type headers had these replaced by the Schmidt type, which became standard on the NER for new and reboilered locomotives.

In LNER days ideas changed, and the

NER Class S2 (LNER Class B15) – Leading details

Designer	Vincent Raven		Grate area	23.00 sq ft
Built	12/1911–3/1913			
Withdrawn	9/1937–12/1947		Tractive effort	21,155 lb
Running numbers	782–799, 813–825			
(original)	(renumbered 1946)		Fuel capacities – coal	5 tons
			– water	3,940 gallons
Wheel diameter				
– coupled	6 ft 1¼ in		Cylinders – number	2 (outside)
– bogie	3 ft 7¼ in		– dimensions	20 in × 26 in
Wheelbase	26 ft 0½ in		Maximum axle load	19 tons 8 cwt
Boiler (as built)			Adhesive weight	53 tons 18 cwt
– diameter	5 ft 6 in			
– length	15 ft 7½ in		Weights in working order	
– tubes, small	268×2 in o/d	(146×2 in o/d)*		Tons Cwt
– tubes, large	–	(24×5¼ in o/d)*	– locomotive	71 02**
			– tender	44 00
Heating surface (as built)			– Total	115 02
– tubes	2153.00 sq ft	(1677.00 sq ft)*		
– firebox	144.00 sq ft	(144.00 sq ft)*		
– Total	2297.00 sq ft	(1821.00 sq ft)*		
– superheater	–	(504.00 sq ft)*		
Working pressure	190 lb/sq in	(160/175 lb/sq in)		

N.B:
* These figures refer to locomotives which were built with superheaters.
** The Stumpf Uniflow locomotive No 825 had the same basic dimensions as the superheated, but was fitted with Walschaerts valve gear, and weighed one ton more.

number of small tubes was increased to 131, and in the early 1930s the abolition of stay tubes increased the number of small tubes to 134. A return was made to the Robinson type header in the superheaters of 10 new boilers, which suffered less from leaks at connections than the Schmidt type headers. These new boilers were built in 1929/1930, and subsequent replacement boilers for this class also contained the Robinson type header.

Design and construction of the smokebox and firebox were similar to the earlier S1 class locomotives. The smokebox was of the cylindrical drumhead pattern, and was enlarged to accommodate the superheater header. On the unsuperheated locomotives, the smokebox was only 3 ft 7¼ in long, the same length as the saddle, while the extension of 4 ft 3½ in in the superheated version overhung the front of the saddle. Originally, a wheel and handle was used to secure the smokebox door, but this was later altered to a twin handle arrangement.

The firebox was a round top design, constructed in the traditional manner with a sheet steel outer casing over a copper inner box, with an overall length of 8 ft 0 in, and providing 140 sq ft of heating surface. The grate, with an area of only 23 sq ft, had a flat rear section over the trailing axle, with the front portion sloping forwards towards the throatplate.

Frames, Wheels and Motion
Main frames were the same as the pioneer class, from mild steel plate, and stayed apart

In LNER livery is Class B15 No 821, with some visible changes from the original, including the provision of pop type safety valves on a large raised mounting on the firebox roof. *L&GRP*

at 4 ft 0 in. Coupled wheels were identical with the S class, at 6 ft 1¼ in diameter, and 3 ft 7¼ in bogie wheels. The main suspension was arranged in the form of helical springs for the leading and driving axles, and leaf suspension for the rear axle. During LNER days, a number of locomotives received solid bronze main axleboxes which appeared in 1934 and 1935 on Nos 787, 797, 799, 823 and 825.

Saturated and superheated locomotives were equipped with 20 in by 26 in cylinders, carrying piston valves above and Stephenson valve gear. As in the S class, the newest NER 4-6-0s had only the cylinders outside the frames, the motion inside, with deeply fluted coupling and connecting rods.

There was one locomotive which was very definitely the odd man out in this class – the first with Stumpf Uniflow (or Unaflow) cylinders to be put to work in this country. The appearance of No 825, which was equipped with the Stumpf system, was quite different. The cylinders with their piston valves above were inclined, but the cylinders were much longer than normal and needed substantially raised footplating to clear them and the Walschaerts valve gear. At the time the Stumpf Uniflow system was popular in Russia, Germany, and France, but this was its first appearance in Britain. In essence, it may have been seen as an alternative to superheating,

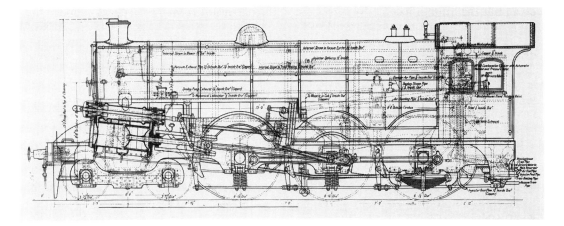

Outline drawing of the NER Stumpf 'Uniflow' locomotive. *The Engineer*

to reduce or eliminate condensation in the cylinders. With the steam travelling in one direction only, there was no rapid heating and cooling of the ports as hot inlet and cooler exhaust gases passed over the port faces. The basic feature of the idea, developed by Professor J. Stumpf of Charlottenburg, was to extract the energy from the steam without causing it to reverse its direction of flow.

In the arrangement adopted by the North Eastern Railway on No 825, the cylinders were still 20 in diameter with a stroke of 26 in, but the overall length of the casting was nearly 5 ft 0 in. There was an inlet port at each end, with a common exhaust port in the centre, consisting of a series of annular holes allowing steam to pass into an exhaust belt surrounding the cylinder. The piston was 2 ft 0 in long, and occupied almost exactly half the cylinder length. In effect this meant that at the end of each stroke one edge or the other was just clearing the ring of holes in the centre, allowing exhaust to take place. The pistons were hollow castings, with spring rings at either end, whose purpose was to act as exhaust valves, uncovering the central exhaust holes as the piston completed 90% of its stroke. The exhaust ports were consequently of fixed size, while the inlet ports, arranged for inside admission, could be varied in the normal way by the driver, according to the cut-off position. The valves had a maximum travel of $6^1/_2$ in, with an exhaust lap of $1^3/_4$ in. When the locomotive was working in full gear, the valve ports acted as auxiliary exhaust, with a small amount of steam exhausted through the inlet ports. However, as the locomotive was notched up, the inlet valves no longer functioned as auxiliary exhaust, with all steam exhausted through the central ring of holes in the centre of the cylinder.

Comparative testing of this locomotive against the conventional design No 797 produced favourable results, even to the extent of fitting one of the NER class Z Atlantics with a similar arrangement. Overall though, the advantages over a conventional superheated locomotive were not sufficient to merit the extended application of the Stumpf system, and No 825 was rebuilt in 1924 as a normal 4-6-0.

Brakes were Westinghouse air-operated for the locomotive, with the compressor carried on the right-hand side next to the firebox; vacuum equipment was provided for passenger train workings. As LNER B15 4-6-0s, these locomotives had their Westinghouse equipment removed in the mid-1930s, to be replaced by Gresham & Craven steam brakes on the locomotive, under the company's unification of brakes programme.

Details and Tenders
Above the footplate the usual commodious two window NER cab was an attractive feature of the design, and while the boiler was larger than in previous types, there was a general family likeness in a number of details.

Narrow wheel splashers, with the tops of the cylinders protruding through the running boards, and a long casing extending from the cab to the cylinders were a fixture on all but the Stumpf locomotive. Over the coupled wheels, extra clearance was allowed in the running boards to accommodate the driving crank. All except No 825 carried piston tail rods. The front footsteps consisted of a single step attached to the rear end of the lower slidebar. Just behind the smokebox mechanical lubricators were carried on both sides for front end lubrication.

Gravity sanding was the order of the day, delivery pipes leading down to the rail behind the trailing and in front of the leading wheels. The sandboxes at the front end were fixed to the frames, behind the slidebars, while the two rear boxes were mounted inside the cab.

On top of the boiler and firebox, the dome housing the regulator was carried on the centre boiler ring, and twin pop safety valves were used from new on all locomotives of the class. The chimney again sported the NER 'windjabber' on the front rim, although over the years a number disappeared as the effects of corrosion were felt.

In LNER days, the B15s received replacement boilers with Robinson type superheaters as standard from around 1931, and all carried the Gresley anti-vacuum valve behind the chimney. There has been the suggestion that only No 786 ran without this

form of superheater protection, although the remaining locomotives were so fitted at the time of their withdrawal.

The tenders paired with these locomotives had a coal capacity of only 5 tons and 3,940 gallons of water, which was carried on a six-wheeled underframe with 3 ft 9$\frac{1}{4}$ in diameter wheels on an equally divided wheelbase of 12 ft 8 in. In the diagrams issued by the LNER for this class, an incorrect water capacity of 4,125 gallons was quoted originally, though this was soon corrected. Three coal rails ran around the tender sides and rear, with a fourth rail added around the coal space. These were later blanked off by fitting sheet steel coping plates behind.

A number of tenders outlived the locomotives to which they were originally attached by many years. Following the withdrawal of the first five B15s in 1937, their tenders were paired with Q6 0-8-0 mineral engines, while the tender of No 813 survived to 1963 paired with a Q6 0-8-0, before it too was scrapped in April that year.

The North Eastern Railway's unique Stumpf 'Uniflow' locomotive, No 825, with its large cylinders on the outside needing a substantially raised footplate. The arrangement and this ungainly looking locomotive did not prove a runaway success, and it remained an isolated example. *National Railway Museum*

Construction and Operation

All twenty S2 (LNER B15) class 4-6-0s were constructed at Darlington between December 1911 and March 1913, and only the first seven were built without superheaters. The LNER issued three diagrams for the class, all Section D, in 1924, 1927, and 1931, and all covering the superheated locomotives, with only minor alterations in successive amendments. No diagrams were issued for either the saturated or Stumpf Uniflow versions. In service after 1923, they were officially described as Southern Area Load Class 3, with a route availability designation of 7.

When first built, the NER allocated its new 4-6-0s to Heaton, Hull (Dairycoates), Leeds (Neville Hill) and York, and they were set to work on genuine mixed-traffic duties, from fish trains to secondary passenger workings. Although their allocations and area of operations remained relatively unchanged at the time of the Grouping, the introduction of more B16 4-6-0s and K3 2-6-0s, relegated the class to goods work. During the 1920s and 1930s the summer timetables saw B15s on passenger work again, mainly on relief trains and excursions to the east coast resorts like Scarborough or Bridlington.

During the 1930s B15s worked regularly over the lines of the former Great Central Railway, on passenger trains between Hull

NER Class S2 (LNER Class B15) 4-6-0 –
Building, renumbering and withdrawal dates

| NER No | Built | LNER Renumbering | | Withdrawn |
		1st 1946	2nd 1946	
782	12/1911	1313	1691	12/1946
786	12/1911			9/1937
787	2/1912	1314	1692	12/1946
788	2/1912			9/1937
791	2/1912			10/1937
795	3/1912	1315		12/1945
796	3/1912	1316		12/1945
797	5/1912			11/1937
798	6/1912	1317		10/1945
799	6/1912			11/1937
813	9/1912	1318		3/1944
815	9/1912	1319	1693*	5/1947
817	10/1912	1320	1694	4/1946
819	10/1912	1321	1695*	12/1946
820	11/1912	1322	1696*	12/1947
821	11/1912	1323	1697*	11/1946
822	11/1912	1324		2/1944
823	12/1912	1325		10/1944
824	12/1912	1326	1698	4/1946
825	3/1913	1327		2/1944

Note:
Of the new running numbers allocated for the 1946 schemes, only those marked* were actually carried.

and Sheffield, while some worked fast goods over the Woodhead route through the Pennines. The later 1930s, which saw the first withdrawals, also witnessed an attempt by the LNER to concentrate locomotive types, particularly at North Eastern Area depots. At the outbreak of World War II Heaton and York had no more B15s, with the remaining 15 allocated to Starbeck, Hull (Dairycoats) and Leeds (Neville Hill). A further concentration moved all the B15s to Hull, where from 1943 they spent the rest of their working lives mainly on goods work, with very occasional passenger turns. The last of the class was withdrawn in 1947.

THE RAVEN S3 CLASS
MIXED-TRAFFIC 4-6-0 –
LNER CLASS B16

The last of the North Eastern Railway 4-6-0 designs which came into the LNER's possession was described as a fast goods locomotive, but was essentially a mixed-traffic type. It was Vincent Raven who saw the culmination of 20 years' development work on the 4-6-0 type in the North East, which had begun with Worsdell's pioneer S class back in 1899. The new 4-6-0s were built from 1919 onwards, and had three-cylinder propulsion in place of the two-cylinder arrangement used previously.

The success of the class Z Atlantics with three-cylinder drive, and LNER's need for more powerful locomotives to deal with the considerable goods and mineral traffic, was pointing the way to three-cylinder propulsion. The first of these more powerful types to appear after World War I was the superheated 0-8-0 mineral engine, from which the same basic design of boiler, cylinders and motion were used for the new S3 4-6-0. In November 1918 an order was placed with Darlington Works for ten 5 ft 8 in 4-6-0s, to be numbered 840–849. So urgent was the need that the first order was not complete before a second order for 25 locomotives was placed in 1919. Less than a year after these were all delivered in June 1921, twenty more were ordered in March 1922, in two batches of ten. December that year saw the final order placed for 15 locomotives, with the last entering service in January 1924.

During Gresley's years as chief mechanical engineer of the LNER, seven of the class

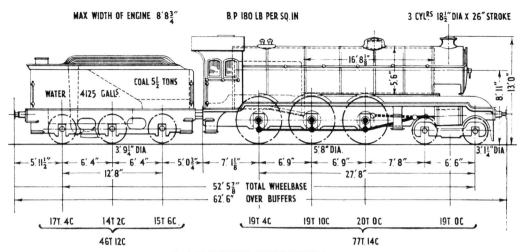

Class B16/1 outline drawing

were substantially rebuilt, as Class B16/2, with Walschaerts valve gear, and derived motion for the inside cylinder, new cabs and a variety of other alterations. Another 17 were rebuilt by Edward Thompson in the late 1940s, but with three separate sets of valve gear, and the same dramatic changes to the appearance of the class that were characteristic of Gresley's rebuilds. The Thompson rebuilds were known as Class B16/3. In their original form, the B16 4-6-0s had the following main dimensions:

Boiler Details

The diagram 49 boilers fitted to these locomotives were the same as those fitted to the T3 (LNER Q7) 0-8-0 mineral engines, and included Schmidt pattern superheaters. The 5 ft 6 in diameter boiler originally housed 102 tubes of 2 in diameter, but their number was increased to 131 in 1927, up to a maximum of 134 in 1933, as spare boilers were built. The latter year marked the beginning of the con-

---

NER Class S3 (LNER Class B16) 4-6-0 – Leading details

Designer	Sir Vincent Raven				Heating surface		
Built	12/1919–1/1924				– tubes	1400.00 sq ft	
Withdrawn	6/1942, 1/1958–7/1964				– firebox	166.00 sq ft	
Original running	840–9, 906/8/9/11/4/5/20–				– Total	1566.00 sq ft	
numbers	34/6/7/42/3, 2362–82, 1371–85				– superheater	392.00 sq ft	
Wheel diameter					Working pressure	180 lb/sq in	
– coupled	5 ft 8 in						
– bogie	3 ft 1 in				Firebox – length	9 ft 0 in	
					– width	3 ft 11 in	
Wheelbase	27 ft 8 in						
					Grate area	27.00 sq ft	
Axleload	*Tons Cwt*	*Tons Cwt*	*Tons Cwt*				
– coupled	20 00	19 10	19 04		Cylinders – number	3	
	Tons Cwt				– dimensions	$18\frac{1}{2}$ in × 26 in	
– bogie	19 00						
					Tractive effort	30,031 lb	
Boiler							
– diameter	5 ft 6 in				Fuel capacities – coal	$5\frac{1}{2}$ tons	
– length	16 ft $8\frac{1}{8}$ in (16 ft $2\frac{5}{8}$ in between tubeplates)				– water	4,125 gallons	
– tubes small	102 × 2 in o/d				Weights (full)	*Tons Cwt*	
– tubes large	24 × $5\frac{1}{4}$ in o/d				– locomotive	77 14	
Superheater elements	24 × $1\frac{3}{32}$ in o/d				– tender	46 12	
					– Total	124 06	

struction of spare boilers, 26 of which had been completed by 1937, to replace the original boilers of 1919, which were becoming life expired. The new replacement boilers included Robinson type superheaters, and with the increase in the number of boiler tubes the heating surface rose from 866 sq ft to 1138 sq ft.

Boilers which had previously housed 102 tubes also included a further seven which had longitudinal stays connecting the firebox and smokebox tubeplates – these were later removed. The later alterations in heating surfaces, and the increase in the number of tubes, was due to re-arrangement of the pattern of installation of the tubes.

The next major change to occur was the construction of the diagram 49A boilers, where the barrel was 6 in shorter and built in two rings where previously there had been three – a sloping firebox throat plate was fitted. The number of small tubes rose to 156, with the heating surface going up to 1293.3 sq ft, and the steam dome was placed further back, to avoid the joint in the two boiler rings. The later diagram 49A boiler was carried by the majority of the B16s, with

Class B16/3 No 61439 at Doncaster shed on 12 April 1959. The Gresley rebuilds of this class added a certain amount of style to their appearance, in the gentle curve of the running boards. This locomotive was rebuilt again by Thompson in 1944, with three sets of Walschaerts gear. *Roger Shenton*

a number of this type being built as replacements in the late 1950s, and pop type safety valves were the standard fitting.

At the leading end, the simple drumhead smokebox was carried on a fabricated saddle, with the chimney sporting the *capuchon*, or 'windjabber,' on the front rim.

The firebox design and construction was traditional, in copper – it measured only 3 ft 11 in wide, with a grate area of 27 sq ft, and remained largely unaltered. The only difference in reboilered locomotives was a slight increase in overall length, by 7/8 in to 9 ft 0⅞ in, with the diagram 49A type, having the sloping throat plate.

Frames, Wheels and Motion
Darlington-built locomotives had frames stayed apart at 4 ft 0 in, which was narrower than contemporary practice generally, including Doncaster, where frame spacing was 4 ft 1½ in. This narrower spacing could have been a factor in the decision not to fit B16s with Thompson's diagram 100A boiler in later LNER days. The cylinders fitted to the Raven 4-6-0s were originally fitted with 8¾ in diameter short-travel piston valves, which under a programme of modification in LNER days were replaced by longer travel valves, 9 in diameter. Changes introduced by Gresley involved new cylinders of the same size as a one-piece casting, with the three sets of Stephenson gear replaced by Gresley's

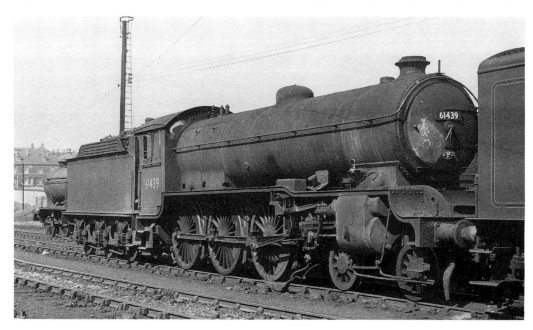

conjugate Walschaerts motion. In order to accommodate these changes the bogie was repositioned 9 in further forward and the locomotives reclassified B16/2. The original NER class S3 was then B16/1. Further rebuilding by Thompson during World War II resulted in the B16/3 class, with three independent sets of Walschaerts gear replacing Gresley's conjugate layout, while the maximum valve travel was raised to $6^3/8$ in, compared to 6 in for the B16/2 version. The original valve travel was a mere $4^7/16$ in. With the last rebuild of the design to B16/3 a change was made to left-hand drive. Out of the total in this class, seven were rebuilt to B16/2, and 17 to B16/3, the former between 1937 and 1940, and the latter between 1944 and 1949.

The locomotive main suspension was by means of underhung leaf springs throughout, with a maximum axle load of 20 tons on the driving axle. The coupled wheels were 5 ft 8 in diameter with 20 spokes, while the bogie wheels were 3 ft 1 in with 12 spokes, in a wheelbase of 6 ft 6 in. Deeply fluted coupling and connecting rods were fitted, and the outside cylinders had tail rods for the pistons, with double slidebars and crossheads carried throughout the locomotives' lives. The running boards were level from the smokebox to just in front of the cab, with a shallow drop at the front end to the buffer beam, and small splashers over the coupled wheels. With the modifications in LNER days, the running boards were raised in a gentle curve over the coupled wheels, as in Gresley's other designs, and splashers were dispensed with on the rebuilds.

When new, all the B16s were fitted with Westinghouse air brakes, and vacuum equipment for working non air-braked trains. From 1928 the LNER introduced a programme to standardise braking systems, with steam brakes on the locomotives and vacuum for the trains. The B16s were dealt with between 1932 and 1936, with minor changes to the layout of pipework and connections.

Details and Tenders
Originally, like their B13 progenitors, the NER's last design of 4-6-0 carried a number of standard features, like the Raven fog signalling apparatus, and steam circulating valves on the superheated locomotives.

Minor details changed under LNER ownership, such as the adoption of standard buffers and drawgear, two-handle fastening of the smokebox door, and steam sanding replacing the compressed-air system used originally.

The locomotives rebuilt to B16/2 and B16/3 had their steam reversing gear replaced by a screw mechanism, while lubrication methods were changed in a number of details during the 1935–1940 period. Originally, the NER adopted mechanical lubrication for cylinders and valves, with the equipment mounted on either side of the locomotive. For the axleboxes, siphon feed was used from oilboxes. The position of the oilboxes feeding the rear journals was changed from the previous location on the cab spectacle plate, and mounted on the firebox side, immediately over the axlebox. Problems of axlebox lubrication at the rear seemed to stem from the position of joints in the ashpan, although no major alterations were made to the lubrication arrangements. Different types of lubricator were tried out, and other minor experimental modifications, but the class as a whole retained the essential features of the NER mechanical lubrication arrangements.

Other minor alterations carried out in LNER days included the fitting of Gresley's anti-vacuum valves, removal of the Raven cab signalling apparatus, the fitting of hinged sight screens between the cab windows. No 1468 was fitted with a B1 type bogie in 1947. The sanding arrangements were altered in some of the rebuilds, notably the B16/3, where rear sanding disappeared for a time. In 1946 No 942 was fitted experimentally with the Downs sanding system. Devised by a running shed foreman in Consett, this equipment, which involved the use of steam heating coils in sandboxes, was also tried on British Railways Standard steam locomotives. The apparatus on the LNER B16s was removed in 1949.

Tenders paired with the B16s were six-wheeled, and carried 4,125 gallons of water, with $5^1/2$ tons of coal in a self-trimming bunker. The tender wheelbase was 12 ft 8 in, equally divided, with conventional outside bearing axleboxes and overhung leaf springs. Four coal rails were fitted, either side of the coal space, with a sheet steel fender behind, and all were fitted with water pick-up gear.

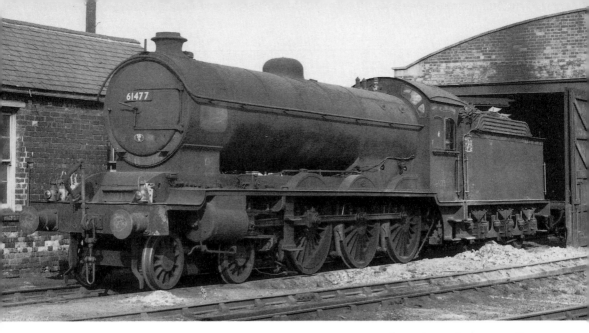

Class B16 No 61477 stands outside York shed on 30 August 1959, one of the many examples remaining in service without the obvious physical changes introduced by Gresley in the 1930s. *Roger Shenton*

No 848 received the tender of withdrawn C7 4-4-2 No 719 in 1944, and was the only recorded case of an exchange of tenders outside the class.

Construction and Operation

All the NER Class S3 4-6-0s (LNER Class B16) were built at Darlington Works between 1919 and 1924. Thirty-two of the 70 locomotives appeared under LNER ownership, coming out between January 1923 and January 1924. Production of the class was quite steady, with five in 1919, nineteen in 1920, eleven in 1921, three in 1922, thirty in 1923, and the last two in January 1924. There was a gap between the appearance of No 943 in June 1921 and No 2363 in November 1922. Not only did the class continue successful careers in LNER days, but all except No 925 became British Railways property in 1948. The missing locomotive was hit during an air raid in World War II while it was stabled at York shed, damaged beyond economic repair, and officially withdrawn in 1942.

The last of the NER's 4-6-0s were to be found at depots close to the East Coast Main Line, with the majority allocated to York in 1924. Some reallocation during the 1920s and early 1930s saw an increase in numbers at Hull (Dairycoates). While these three-cylinder locomotives have been referred to as express goods engines, they were often seen as mixed-traffic types, working relief and excursion passenger trains. Freight work involved some very heavy haulage in the 1920s and 1930s, particularly for the Leeds, Hull and York based locomotives. Just about every conceivable kind of goods train was handled by B16s, although on passenger duties the class seldom handled the heaviest main line expresses.

In the mid 1930s, the class had settled in greater numbers at Neville Hill, Dairycoates and York, where the allocations amounted to 9, 26 and 22 respectively – the remaining 23 were shared between five depots in the North East. Some B16s had worked north of the border to Edinburgh, while York based examples went as far as Sheffield, Manchester, and Nottingham. World War II saw a concentration of the remaining 69 locomotives in the York area, working all types of train, and there they remained during the final years of LNER ownership, extending their routes further south over Great Central metals, even into Marylebone. They were more common on the Great Northern line in the 1940s, substituting for B1s, and V2s, and working into King's Cross on a York V2 diagram. Further service on important passenger workings saw B16 power during BR days, although in 1946 No 1458 had already worked the 7.05pm down relief Aberdonian. The class was to last another 16 years in British Railways ownership until the mid 1960s, performing successfully on demanding passenger workings, better in some cases than some of their descendant designs.

The Raven NER Class S3 (LNER B16) 4-6-0 – Building and withdrawal

Original Nos	LNER 1946 Nos	BR Nos	Built	Withdrawn	Scrapped
840	1400	61469*	12/1919	10/1960	Darlington
841	1401	61470*	12/1919	11/1959	Darlington
842	1402	61471*	12/1919	9/1960	Darlington
843	1403	61472*	12/1919	4/1964	Draper, Hull
844	1404	61473*	12/1919	9/1961	Darlington
845	1405	61474*	4/1920	1/1958	Darlington
846	1406	61475*	3/1920	4/1963	Darlington
847	1407	61476*	3/1920	9/1963	Darlington
848	1408	61477*	4/1920	2/1960	Darlington
849	1409	61478*	4/1920	12/1960	Darlington
906	1410	61410	6/1920	10/1960	Darlington
908	1411	61411	8/1920	9/1961	Darlington
909	1412	61412	8/1920	9/1961	Darlington
911	1413	61413	8/1920	9/1961	Darlington
914	1414	61414	9/1920	9/1961	Darlington
915	1415	61415	9/1920	9/1961	Darlington
920	1416	61416	9/1920	5/1961	Darlington
921	1417	61417	10/1920	9/1962	Darlington
922	1418	61418	11/1920	6/1964	Draper, Hull
923	1419	61419	11/1920	9/1961	Darlington
924	1420	61420	12/1920	6/1963	Draper, Hull
925	—	—	12/1920	6/1942	
926	1421	61421	12/1920	6/1964	Hughes Bolckowe Ltd
927	1422	61422	12/1920	9/1961	Darlington
928	1423	61423	1/1921	9/1961	Darlington
929	1424	61424	2/1921	10/1960	Darlington
930	1425	61425	2/1921	9/1961	Darlington
931	1426	61426	3/1921	9/1959	Darlington
932	1427	61427	3/1921	3/1960	Darlington
933	1428	61428	3/1921	10/1960	Darlington
934	1429	61429	4/1921	9/1961	Darlington
936	1430	61430	4/1921	10/1959	Darlington
937	1431	61431	5/1921	9/1961	Darlington
942	1432	61432	6/1921	7/1961	Darlington
943	1433	61433	6/1921	11/1959	Darlington
2363	1434	61434	11/1922	6/1964	Hughes Bolckowe Ltd
2364	1435	61435	11/1922	7/1964	Draper, Hull
2365	1436	61436	12/1922	9/1961	Darlington
2366	1437	61437	1/1923	6/1964	Draper, Hull
2367	1438	61438	1/1923	6/1964	Draper, Hull
2368	1439	61439	2/1923	8/1962	Darlington
2369	1440	61440	3/1923	8/1960	Darlington
2370	1441	61441	3/1923	10/1959	Darlington
2371	1442	61442	4/1923	2/1960	Darlington
2372	1443	61443	4/1923	9/1961	Darlington
2373	1444	61444	5/1923	6/1964	Draper, Hull
2374	1445	61445	5/1923	7/1961	Darlington
2375	1446	61446	5/1923	1/1961	Darlington
2376	1447	61447	6/1923	9/1961	Darlington
2377	1448	61448	5/1923	6/1964	Hughes Bolckowe Ltd
2378	1449	61449	7/1923	7/1963	Darlington
2379	1450	61450	6/1923	9/1961	Darlington
2380	1451	61451	7/1923	9/1961	Darlington
2381	1452	61452	8/1923	9/1961	Darlington
2382	1453	61453	8/1923	6/1963	Darlington
1371	1454	61454	10/1923	6/1964	Hughes Bolckowe Ltd
1372	1455	61455	10/1923	9/1963	Darlington
1373	1456	61456	10/1923	8/1960	Darlington
1374	1457	61457	10/1923	6/1964	Hughes Bolckowe Ltd
1375	1458	61458	10/1923	11/1959	Darlington
1376	1459	61459	11/1923	9/1961	Darlington
1377	1460	61460	11/1923	9/1961	Darlington
1378	1461	61461	11/1923	9/1963	Darlington
1379	1462	61462	11/1923	5/1961	Darlington
1380	1463	61463	12/1923	6/1964	Draper, Hull
1381	1464	61464	12/1923	9/1963	Darlington
1382	1465	61465	12/1923	1/1960	Darlington
1383	1466	61466	12/1923	7/1961	Darlington
1384	1467	61467	1/1924	6/1964	Draper, Hull
1385	1468	61468	1/1924	9/1963	Darlington

Notes:
All locomotives were built at Darlington
* These locomotives were originally numbered 61400–61409 by British Railways, although Nos 61403/8/9 never carried
  them.

CHAPTER THREE
THE ROBINSON FAMILY

Under the authority of John Robinson, the Great Central Railway was not slow to produce locomotives with the 4-6-0 wheel arrangement. Between 1902 and 1922 no fewer than nine different classes appeared on that railway, with the Immingham and Sir Sam Fay classes being perhaps the most well known. The earliest 4-6-0s built by the Great Central were almost the most numerous – the Class 8, which totalled 14 locomotives, survived from 1902 until early British Railways days.

Although many of the GCR locomotives were built at the company's Gorton Works in Manchester, some were constructed by Beyer Peacock, and Neilson & Co. The last of Robinson's designs included ten that were built by the LNER, after the 1923 Grouping, with a number emerging from the Vulcan Foundry Newton-le-Willows works. Traffic

The massive-looking No 423 *Sir Sam Fay* at Neasden, still in GCR colours, in May 1923 – one of the few inside-cylinder 4-6-0 locomotives possessed by the GCR. *L&GRP*

handled by the Robinson locomotives ranged from goods, mixed-traffic, to the most prestigious passenger workings. However, in service, a number of types earned a reputation as excessive consumers of coal, and displayed shortcomings in design and construction.

When in the 1930s the Gresley Sandringham 4-6-0s arrived on Great Central metals, there was some displacement from passenger duties, and other principal workings handled by the Robinson locomotives. After World War II developments in locomotive design policy, operating practices, shortage of labour, and nationalisation, resulted in their fairly rapid demise. The last of the Robinson types was withdrawn in 1950. During their years of service, in addition to the introduction of new technology – superheating, for instance – some locomotives received more extensive modifications, such as the fitting of poppet valve gear, although these alterations were less extensive in many instances than had occurred to other companies' designs, during LNER service.

The Left-hand side of one of the most numerous of ex-GCR types, the Class 9Q, LNER Class B7, 'Black Pigs', still sporting the Robinson top feed, which later disappeared, along with the steam recirculating valves seen on the smokebox side. *Lens of Sutton*

The number and haulage capacity of the 4-6-0 types built for the Great Central was consistent with the company's attempts to secure a share of the passenger market on its recently completed London extension. During the 1920s and 1930s, after the take-over by the LNER, competition for express passenger traffic into the Marylebone terminus was maintained, despite the popularity and tradition of service into both East Coast and West Coast main line termini. The substantial goods traffic was really the mainstay over the GCR route, and working five-coach passenger trains with locomotives of the Sir Sam Fay, or Immingham classes was an extravagant use of power.

Transfers to the former Great Northern main line early in the 1920s resulted in six out of the nine former GCR types putting in appearances in King's Cross by 1926. Two classes – which eventually became LNER Classes B6 and B9 – did not work south of Peterborough, although members of the former class could be found at ex-GNR sheds elsewhere on the system.

An interesting series of trials involving the B3 and B7 classes – nicknamed 'Black Pigs' – took place over the Woodhead route from Sheffield into Manchester in 1927. These two types together with a D11 class 4-4-0 were used to provide information on the traction needs for electric locomotives on this steeply-graded route over the Pennines. These tests were carried out some years before authority for electrification was given, while in 1948 further trials of steam types

were held over the route under the auspices of British Railways.

The Robinson 4-6-0s like other pre-Grouping designs, survived for longer under Gresley's motive power policies than they might have done on (say) the LMS. The series of ex-GCR types were maintained in traffic, with some alterations; had not World War II intervened, and with Edward Thompson as chief mechanical engineer they might have disappeared more quickly.

THE PASSENGER CLASSES

Four different classes of passenger locomotive were built to Robinson's design and instructions between 1903 and 1917, totalling 24 locomotives, half of which were constructed by Beyer Peacock. The first of the new 4-6-0 types was classified 8C by the Great Central, becoming B1 (later B18) under LNER ownership. The two locomotives were in the nature of an experiment, to compare the performance and suitability of the 4-6-0 and 4-4-2 wheel arrangements for passenger service. The two class 8C 4-6-0s were ordered along with two 4-4-2 types, with a number of dimensional similarities. Both designs were two-cylinder simple engines, and the 4-6-0s were given running numbers 195 and 196.

GCR Class 8C (LNER Class B1–later B18) 4-6-0 – Building and withdrawal

GCR No	Built	Works No	LNER 1924 No	1st 1946 No	2nd 1946 No	Withdrawn
195	12/03	4541	5195	1470	1479	12/47
196	1/04	4542	5196	1471	1480	12/47

Both locomotives were built by Beyer Peacock.

Construction

The new 4-6-0s were certainly impressive looking machines, with 6 ft 9 in coupled wheels, 19½ in cylinders in No 196, and 19 in diameter in No 195. The cylinders were outside the frames, with their short-travel slide valves inside, actuated by Stephenson valve gear. With a total length of almost 62 ft 0 in, the new locomotives tipped the scales in working order at 107 tons.

Initially Nos 195 and 196 were not equipped with superheaters. The boiler, built up from three rings, was 15 ft 0 in long, and pitched at 8 ft 6 in above rail level, with a working pressure of 180 lb/sq in. A large round-topped steam dome was carried on the centre ring. The boiler housed 221 2in diameter tubes, giving a heating surface of 1777.9 sq ft. The firebox was of the Belpaire type, waisted-in to sit between the frames, and providing 133.1 sq ft of heating surface. The inner firebox was copper, measuring 7 ft 9¹¹/₃₂ in by 3 ft 4¹/₈ in and had a grate area of only 26 sq ft. Considering the general size of the locomotive, this seems rather small, and in later comparative trials against the ex-GNR Atlantics, GCR 4-6-0s were unable to match the performance of the 4-4-2s despite the advantage of the wheel arrangement. Later modifications included the fitting of superheaters, with Robinson's modifications to the original Schmidt pattern including the expansion of the ends of the elements into the headers, in order to eliminate leakage from pipework joints. The header discharge valves fitted to the sides of the smokebox were subsequently replaced by the Gresley design of snifting valve, usually fitted behind the chimney.

The locomotives included more extensive use of steel castings in their construction than previous practice, and the cylinders in each of the two 4-6-0s were of two different diameters, 19 in and 19½ in, with a common stroke of 26 in. A similar arrangement was adopted for the Atlantic types. The cylinders were inclined at an angle of 1 in 48, and drove onto the centre pair of coupled wheels. Balanced slide valves were used in these first 4-6-0s, with a maximum travel of only 4¼ in, although in later designs piston valves were preferred. The main frames were 1¼ in thick, stayed apart at 4 ft 1½ in, but narrowed in to 3 ft 10½ in at the cylinders, and 3 ft 7½ in at the front buffer beam. Coupled wheels of 6 ft 9 in diameter – in common with the Atlantics – with 3 ft 6 in bogies wheels, were included in a total wheelbase of 27 ft 9½ in.

The main suspension utilised leaf springs for the leading and trailing coupled axles and bogie, with coil springs for the driving axle. Steam sanding was fitted ahead of the leading wheels. Steam brakes were provided for locomotive and tender, with vacuum brakes for the train. The tender, weighing-in at 39 tons 6 cwt, had a six-wheel underframe with sheelbase of 13 ft 0 in equally divided; it carried 5 tons of coal, and 3,250 gallons of water.

Both locomotives were built by Beyer Peacock in 1903, and classified 8C by the Great Central; under LNER ownership they became class B1 (later B18) and carried running numbers 5195 and 5196. Both locomotives were LNER lined green livery, although some other GCR types were treated to a lined black colour scheme. The only modification to these 4-6-0s in LNER days was the fitting of superheaters.

The 8F Immingham Class 4-6-0 (LNER Class B4)

The next generation of 4-6-0s to be put to work on the Great Central for passenger duties were the members of Class 8F, later classified B4 by the LNER. There were 10 locomotives in this class, all of which survived into LNER ownership, with four of their number still at work in 1948.

GCR Class 8F (LNER Class B4) 4-6-0 – Building and withdrawal

GCR No	Built	BP Works No	1924 No	1st 1946 No	2nd 1946 No	BR No	Withdrawn
1095	6/06	4816	6095	1490	—	—	2/44
1096	6/06	4817	6096	1491	1481	—	7/47
1097	6/06	4818	6097	1492	1482	61482	11/50
1098	6/06	4819	6098	1493	1483	61483	9/49
1099	6/06	4820	6099	1494	1484	—	11/47
1100	6/06	4821	6100	1495	1485	61485	6/49
1101	6/06	4822	6101	1496	1486	—	10/47
1102	6/06	4823	6102	1497	1487	—	12/47
1103	7/06	4824	6103	1498	1488	61488	10/48
1104	7/06	4825	6104	1499	1489	—	7/47

Notes:
All locomotives built by Beyer Peacock.
No 1097 was named *Immingham*

The new locomotives in the Robinson stable were intended for express passenger and excursion work, but which could if the need arose be employed on fast freight services. The contemporary press referred to their particular use on non-stop trains with heavy loads. The third locomotive in the class, No 1097, was given the name *Immingham* to mark the beginning of construction work on the company's new docks at that port. In July 1906 No 1097 was used to haul the company's directors from London to Immingham for the ceremonies.

In the design, however, the only real departure from previous types was in the size of the coupled wheels, as comparison of the leading dimensions of the first two types reveals. (See tables on pages 47 and 48.)

Construction
Like its progenitors, the Immingham type had a boiler constructed in three rings. The boiler was fast becoming a GCR standard type, and was also used on the Robinson 2-8-0 goods locomotives, which became more familiar later as the ROD type. The standard assembly consisted of parallel boiler, Belpaire firebox, and originally a short smokebox, since the locomotives were not superheated.

Below the footplate the two cylinders were 19¼ in diameter, outside the frames, with steam chests and valves inside and operated by Stephenson valve motion. Balanced slide valves were used once again, although under Gresley piston valves were substituted.

Weight distribution on the 6 ft 7 in coupled wheels was fairly even, with 18 tons 6 cwt on the leading and driving wheels, and 17 tons 18 cwt on the trailing pair, in a total wheelbase of 26 ft 9½ in. Locomotive main and bogie suspension incorporated leaf springs. Sandpipes were arranged to deliver in front of the leading wheels only, as in the first GCR 4-6-0 design. Braking equipment was also the same as used on Nos 5195 and 5196.

The cylinders included long tail rods for the pistons, with double slidebars fixed to the back cylinder cover and their outer ends fixed to a motion bracket attached to the frames in front of the leading coupled wheels. Screw reverse was used, with the reach rod attached to the top of a pivoted arm on the footplate at its leading end between the coupled wheel splashers. A shorter rod was then employed, extending forward to raise and lower the link.

Boiler fittings were similar to the earlier design with the large round-topped steam dome on the middle boiler ring, and typical Robinson tapered chimney surmounting the smokebox. Initially, four Ramsbottom type safety valves were carried on the firebox roof, in a brass casing. The working pressure of these boilers has been quoted as 180 lb/sq in, but some contemporary sources refer to a figure of 200 lb/sq in. Similarly, the coupled wheel size is frequently stated to be 6 ft 7 in, while 6 ft 6 in has been mentioned, again in contemporary sources.

The tenders for the new Immingham class were similar to their predecessors but weighed two tons more, carrying five tons of coal, and 4,000 gallons of water. Overhung leaf springs to the outside bearing axleboxes supported the total weight of 41 tons 3 cwt. A minor detail alteration was the inclusion of sheet steel fenders at the sides of the coal space, in preference to the three coal rails used previously.

The Class 1 Sir Sam Fay 4-6-0 (LNER Class B2)

Third in the quartet of GCR passenger 4-6-0s were the six inside-cylinder locomotives of Class 1, later to become LNER Class B2. Built at the company's Gorton Works, they were much larger than previous designs, with the purpose of accommodating the 'rapidly increasing speed and weight of trains.' The services on which it was intended to employ the new class included the heaviest express passenger and mail trains, excursions, fish, and express goods workings.

Only six of the class were built, in 1912–1913, with the first locomotive, No 423, despatched for exhibition at Ghent in Belgium in 1913. An interesting claim for the new locomotives – the first to be built with the Robinson pattern superheater – was that they were 50% more powerful than the company's Atlantic and existing 4-6-0 designs. With such a large boiler, the narrow firebox with its equally narrow grate, having an area of only 26 sq ft, was less than adequate. In comparative trials with former GNR Atlantics with their wide firegrates the adage, often ascribed to H. A. Ivatt, that a measure of a locomotive's performance lay in its ability to boil water, certainly proved true in this case. For certain workings into Leeds, on which GCR 4-6-0s were tried, the GNR Atlantics were preferred.

On Britain's railways up to about 1920 and for some years afterwards the 4-6-0 type seemed to take a long time achieving its potential, except perhaps on the GWR. For mixed-traffic purposes, it was probably the best choice, but in the early developments it was in the role of passenger service that its characteristics were being harnessed. Having said that, out of the nine Robinson designs, only four were passenger types, the most numerous and successful of the company's designs was a mixed-traffic class.

The Sir Sam Fay class was designed up to the limits of the GCR loading and structure gauges, turning the scales at 122 tons 11 cwt, it would still fit onto a 55 ft turntable.

Boiler Design

The design which emerged from Gorton carried a new and much larger boiler than previous types, being superheated from the start. The same basic construction methods were used as previously, resulting in a boiler 5 ft 6 in diameter, and 17 ft 3 in between tubeplates. Housed within the boiler originally there were 157 $2^{1}/_{4}$ in diameter tubes, and 24 $5^{1}/_{4}$ in diameter superheated flues incorporating 24 $1^{3}/_{8}$ in elements. However the boiler despite its size was inadequate to meet the demands of the $21^{1}/_{2}$ in diameter cylinders, and the first modifications included a reduction in the number of small tubes to 139, and superheater elements to $1^{1}/_{16}$ in diameter. Final alterations to the boiler included 28 superheater flues, and a further reduction in the number of small tubes to 116; in this guise the boiler became diagram 13 under the LNER classification. With the heating surface down to 1203 sq ft from 1630 sq ft, and the superheater heating surface down to 343 sq ft from 440 sq ft some success was achieved, although still not entirely satisfactorily, as the cylinders of Nos 423/4/7

GCR Class 1 (LNER Class B2) – Building and withdrawal

GCR No	Built	1924 No	Name	1st 1946 No	2nd 1946 No	Withdrawn
423	12/12	5423	Sir Sam Fay	1472	1490	4/47
424	1/13	5424	City of Lincoln	1473	—	11/45
425	2/13	5425	City of Manchester	1474	1491	7/47
426	3/13	5426	City of Chester	1475	—	12/44
427	3/13	5427	City of London	1476	1492	11/47
428	12/13	5428	City of Liverpool	1477	1493	4/47

Notes:
1. All locomotives built at Gorton Works
2. No works numbers issued
3. *City of London* name removed September 1937 — name transferred to Class B17 No 2870.

were lined-up to 20 in diameter in 1921/1922. The LNER reclassified the B2s with 20 in cylinders B2/2, and the 21½ in diameter cylinders were found on class B2/1. The locomotives with the original 24-element superheater survived in this form until the mid 1930s, when No 5427 was reportedly the last to be fitted with this boiler.

Although the superheaters fitted were basically the Schmidt pattern, Robinson incorporated significant design improvements, and the arrangement was adopted as standard by the LNER. In addition to GCR types it was fitted retrospectively to some NER locomotives. Control of the installation of Robinson superheaters was vested in the Superheater Corporation Ltd. In 1915, following the successful application in rail service, a marine version was announced, the technicalities of which attracted some interest in the contemporary press.

Connection of superheater elements with the smokebox header had previously required the use of glands, which were inevitably susceptible to leakage. Robinson's main change was to expand the ends of the elements directly into the header. With the Sir Sam Fay class originally twenty-four 1⅜ in elements were provided, although as already mentioned this was increased even before the Grouping with elements of reduced diameter, in an effort to improve the performance of this boiler.

The damper mechanism, with its exterior operating handle on the side of the smokebox (a common feature of a number of contemporary designs) was also eliminated in an ingenious way. To prevent the possibility of burning or overheating elements when no steam was present, as when the regulator was closed, a small jet of steam to each of the flues, controlled by use of the blower, counteracted the draught through the large tubes. This equipment was later removed, and replaced during GCR days by another Robinson idea – a combined blower/circulating valve – whose function, governed by the opening of the regulator and blower, admitted steam to the superheater elements in small quantities. In order to prevent movement of the locomotive with the regulator closed but when steam was still passing through the superheater elements header discharge valves were fitted, and were a noticeable feature on the smokebox.

At the opposite end of the boiler the Belpaire firebox, constructed from steel plate for the outer box and copper for the inner, provided 167 sq ft of heating surface. Externally, it was 8 ft 6 in long but shallow, with its meagre 26 sq ft of grate area and paltry 7 in deep ashpan carried over the rear coupled axle. This inadequate arrangement was the same as that fitted to the first pair of Robinson 4-6-0s; it did not include a rear damper, but a rather high firehole.

Mechanical Features
The real shortcomings in the new design appeared below the footplate, where the outside cylinders had given way to two very large inside cylinders, while the slide valves had been replaced by piston valves, but still employing a short travel. The cylinders were 21½ in diameter originally; their correspondingly increased centre-to-centre distances prevented the fitting of more generously proportioned axleboxes, with the occurrence of overheating due to the restricted clearances. The 10 in diameter piston valves on top of the cylinders were arranged for inside admission, and operated by rocking shafts from the Stephenson valve motion. Here too there were design shortcomings that affected performance, since the heads of the valves were closer together than usual – this restricted the layout of the steam circuit, making the passages more tortuous. The cylinders were fitted with wrought-iron pistons, and Robinson's pressure relief valves, taking the place of the air and bypass valves more usually seen on locomotives of the day fitted with this type of piston valve. The first five members of the class had connecting rods with marine type big ends, 6 ft 6 in long, while the last member of the class was equipped with a more conventional design. The large diameter of the cylinders was the main factor in the poor performance of these locomotives, and resulted in major alterations to the layout of the boiler, by both the Great Central and LNER, the latter going to the extent of lining-up the cylinders of Nos 5423/4/7 to 20 in.

Engineering reports of the day suggest that the 6 ft 9 in diameter coupled wheels were provided with generously proportioned coupled axleboxes. For the trailing and centre pairs these were 8 in diameter and 10 in long, but for the driving pair, although they were 9 in diameter, they were also only 9 in long. In view of the operating problems encountered, these dimensions were clearly not generous enough. Suspension was a combination of leaf and helical springs, with the latter used in pairs on the driving and centre coupled axles, and laminated springs on the rear. The conventionally built plate frame bogie, with its 3 ft 6 in wheels also had leaf springs for its main suspension, with the $6^1/_2$ in sideplay controlled by a laminated elliptic spring.

Detail alterations before the Grouping included the replacement of the mechanical lubricators with Robinson's Intensifore system, ash ejectors were installed, and the four Ramsbottom safety valves gave way to Ross pop type. An interesting modification was the inclusion of the Reliostop system of automatic train control on No 423 in 1919 – the apparatus was removed two years later.

In 1921 oil burning equipment was fitted to five of the class. The conversions were made as a result of disputes in the coal industry, and similar reasons resulted in its re-installation by the LNER in 1926. On both occasions the installations barely lasted six months.

No 1169 *Lord Faringdon* on a down Manchester train, near Ashby in 1921, hauling an interesting collection of coaches. *L&GRP*

Footplating was raised over the coupled wheel crankpins, with a single large splasher over the wheels carrying the locomotives' names. A conventional Robinson type cab was fitted, although the dome casing had a flatter top, and the chimney was slightly more squat than on previous designs. Improvements in sanding arrangements provided sandpipes leading down to rail level in front of the leading and intermediate wheels, and to the rear of the trailing pair. Although the locomotives were not overtly successful in performance, and may well have resulted in the appearance of the eminently popular Director class 4-4-0s, paired with their 4,000-gallon capacity tenders, they were certainly an attractive design.

Lord Faringdon – The last passenger 4-6-0
In 1917 the last Great Central contribution to the LNER stock of passenger 4-6-0s emerged from Gorton Works, in the shape of six locomotives classified 9P by the Great Central (LNER B3). Locomotives of this class were subjected to some interesting trials during LNER days, and a number were fitted with Caprotti valve gear in the 1930s.

Once again, the Great Central had produced a large 4-6-0 type for heavy passenger working, along with some equally heavy fish workings between Grimsby and Maryle-

GCR Class 9P (LNER Class B3) — *Building and withdrawal*

GCR No	Built	1924 No	Name	1st 1946 No	2nd 1946 No	BR No	Withdrawn
1169	11/17	6169	Lord Faringdon	1480	1494	—	12/47
1164	6/20	6164	Earl Beatty	1481	1495	—	9/47
1165	7/20	6165	Valour	1482	1496	—	12/47
1166	8/20	6166	Earl Haig	1483	1497	61497	4/49
1167	9/30	6167	Lloyd George	1484	1498	—	12/47
1168	10/20	6168	Lord Stuart of Wortley	1485	1499	—	9/46

Notes:
All locomotives built at Gorton. No works numbers issued.
Names removed: No 1497 October 1943, No 1167 (1497) August 1923.

bone, but again there were some short-comings in the design. The basic layout incorporated the same type of boiler as used in the previous design, with a standard GCR 4,000-gallon tender, but this time with four cylinders driving the 6 ft 9 in coupled wheels. A couple of years later, the same layout was used for a goods version, which had acquired the unenviable nickname 'Black Pigs' due to their voracious appetite for coal. Despite this, these latter locomotives formed the GCR's most numerous design. The six locomotives of the Lord Faringdon class were given running numbers 1164–1169.

Construction

The boilers were the same as for the Sir Sam Fay class, 5 ft 6 in diameter and 17 ft 3 in long. Changes were restricted mainly to the fitting of Robinson pattern superheaters from new, with 28 short return elements, $1^{1}/_{16}$ in diameter in $5^{1}/_{4}$ in diameter flues.

There was one exception however: No 1169 seems to have been fitted with the 24-element superheater, as in the second modification to the B2 boiler. As LNER No 6169, this locomotive was brought into line with the rest in December 1923. The original heating surface of the 28-element superheater was 343 sq ft, with the 116 $2^{1}/_{4}$ in diameter tubes providing 1,881 sq ft of evaporative heating surface, and another 163 sq ft from the firebox. Robinson's steam circulating valves and header discharge valves were fitted to all from new. Again, the major shortcoming of the B3 class was the inadequate 26 sq ft of grate area.

Beneath the footplate, the new 4-6-0 was a very different animal indeed, with Robinson adopting four 16 in diameter cylinders, in line under the smokebox, with divided drive. Stephenson valve gear was used, in one of the

Outline diagram of GCR Class 9B (LNER B3)

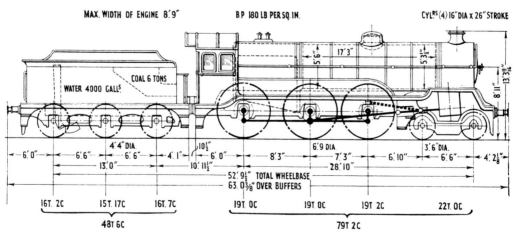

few applications in a four-cylinder type, in common with the 'Black Pigs.' The inside cylinders drove onto the leading axle, with the outside cylinders driving the centre pair, which arrangement required longer than normal connecting rods outside. The 8 in diameter piston valves had the short travel of only 4³/₈ in, and were driven from the same set of motion by means of rocking shafts. Inside admission used for the inside valves, with outside admission for the outside pair.

The cranks for the outside cylinders were set at 180 degrees to one another, with those of the two inside cylinders then set at 90 degrees to each other, to avoid any problems with dead centre positioning. For the outside cylinders, the slidebars were incorporated in a large casting attached to the main frame, instead of the more usual practice of attachment to the back cylinder cover. The result of this was longer than normal piston rods, a feature also seen on the other Robinson four-cylinder 4-6-0 design.

The installation of poppet valve gears was intended to make steam locomotives more economical in service, and with the B3s' reputation for fuel consumption the LNER selected Nos 6166 and 6168 in 1929 for conversion. The Caprotti gear fitted was manufactured by Beardmore under licence to Italian design, with two more sets made by Armstrong Whitworth & Co to the British design of Caprotti Valve Gears Ltd, and fitted to Nos 6164 and 6167. In 1929, the cost of the two camboxes, drive and reversing mechanism was £2,211-2s-0d [£2,211.10].

On the B3 class, one vertically mounted poppet valve was provided for inlet and one for exhaust, at each end of the cylinder, with a rotary drive from the driving axle. The transverse shaft mounted across the cylinders was driven by gearing, with cams on the shaft operating the inlet and exhaust valves by a series of bell cranks. In the first two conversions springs were used to re-seat the valves, whereas in the later British designs supplied for Nos 6164/6167 in 1938/39 an auxiliary steam supply was used. Although the equipment was supplied for the second pair of locomotives the conversion was not completed and they ran with Stephenson gear throughout their careers.

The 6 ft 9 in coupled wheels were carried in 8 in diameter by 9 in long journals on the crank axle, and 8 in diameter by 12 in long journals on the two remaining axles. Axlebox lubrication pads were carried in spring-supported cast-iron keeps, an advantage of which was to enable their removal without the need to take down the spring gear. Locomotive main and bogie suspension was laminated springs, with the side play in the bogie also controlled by leaf springs, in what was GCR standard practice. Automatic steam brakes were provided for the locomotives, with vacuum equipment for the train.

Running boards were raised at the front end over the cylinders. A single splasher covered the coupled wheels and made a suitable location for nameplates. On the Caprotti-fitted locomotives a large box with hinged access doors covered the camboxes and valves, while on all locomotives the pipes for the smokebox ash ejectors could be seen on the outside of the smokebox.

Originally, the firebox roof on the first of the class sported Ramsbottom type safety valves, which were later replaced with a pair of Ross pop valves, in common with the rest of the class.

The cab, large as it was on most ex-GCR types, with the prominent roof ribs, was altered in three examples to have two side windows instead of the single cut-away in the side sheets. Nos 1165, 1167 and 1168 had this modified cab. Tenders paired with the B3s were Robinson's standard six-wheel type with space for six to seven tons of coal and 4,000 gallons of water.

FREIGHT TYPES

The 'Fish Engines'

The Great Central built only two freight type 4-6-0s, which later became LNER stock, the first of these being the Class 8 (LNER B5) design of 1902. A second type, Class 8G (LNER B9) appeared from 1907.

The Class 8 was the earliest 4-6-0 design to enter service with the Great Central, with 6 ft 1 in diameter coupled wheels, and saturated boilers pressed to 180 lb/sq in. They were built in two batches, with the first six coming from Neilson and Co, and the last eight from Beyer Peacock two years later, in 1904. Intended for working the Grimsby fish trains they acquired the nickname 'Fish Engines' although in later years they were also employed on some passenger workings.

GCR Class 8 (LNER Class B5) 4-6-0 – Building and withdrawal

GCR No	Built	Builder	Works No	1924 No	1st 1946 No	2nd 1946 No	BR No	Withdrawn
1067	11/02	Neilson & Co	6235	6067	1300	1678	—	11/47
1068	12/02	Neilson & Co	6236	6068	1301	1679	—	12/47
1069	12/02	Neilson & Co	6237	6069	1302	1680	61680	11/48
1070	12/02	Neilson & Co	6238	6070	—	—	—	3/39
1071	12/02	Neilson & Co	6239	6071	1303	1681	61681	6/48
1072	12/02	Neilson & Co	6240	6072	1304	1682	—	8/47
180	1/04	Beyer Peacock & Co	4531	5180	1305	1683	—	12/47
181	1/04	Beyer Peacock & Co	4532	5181	1306	1684	—	5/47
182	1/04	Beyer Peacock & Co	4533	5182	1307	1685	61685	3/48
183	2/04	Beyer Peacock & Co	4534	5183	1308	1686	61686	6/50
184	2/04	Beyer Peacock & Co	4535	5184	1309	1687	—	7/47
185	2/04	Beyer Peacock & Co	4536	5185	1310	1688	61688	11/49
186	2/04	Beyer Peacock & Co	4537	5186	1311	1689	61689	10/49
187	3/04	Beyer Peacock & Co	4538	5187	1312	1690	61690	4/48

Their original saturated boilers remained throughout their Great Central careers, and it was not until after they became LNER Class B5 that any modifications were made. The original boilers were 4 ft 9 in diameter and 15 ft 0 in long, housing 207 small 2 in diameter tubes with a heating surface of 1665.0 sq ft. Between 1923 and 1936 all the B5s were fitted with superheated boilers. No 184 (later 5184), the first to be altered, and was given a 5ft 0 in diameter boiler – diagram 15 in the LNER classification – and the same as that carried by the O4 class 2-8-0s. The remainder of the Fish Engines had the diagram 17 boilers, as fitted to Q4 class 0-8-0s, and in later years even the first superheated B5 was fitted with a diagram 17 boiler. Their

rebuilding/reboilering in LNER days resulted in sub-divisions of the class, with the unrebuilt locomotives as B5/1. The first rebuilding of No 5184 became class B5/2, with the second reboilering becoming B5/3. This latter was the LNER classification for all the superheated Fish Engines, and the sub-divisions were abolished in 1937.

Nos 6067, 6072, 5180, 5183, 5184, 5185 and 5187 were fitted with new 21 in diameter O4 class cylinders, with 10 in diameter piston valves, while the other half of the class retained their original slide valves and 19 in

One of the Class B5 4-6-0s, No 1068, on a train of fish empties near Rothley in 1922. Note the casing round the safety valves on the Belpaire topped firebox. *L&GRP*

by 26 in cylinders. Some minor changes were made to boiler fittings so that the locomotives would meet standard LNER loading gauge dimensions. Other minor alterations by the company ranged from removal of the piston tail rods to different methods of fastening the smokebox door. Live steam injectors replaced the live and exhaust steam combination of earlier years on some locomotives, while the ash ejectors were removed during World War II. An interesting experimental fitting on this class was the Schleyder ash consumer on No 1072 in 1912. With this equipment, ashes were sucked from the smokebox and passed back into the firebox, for a second time to ensure complete combustion of any partially-burned matter. The equipment was later removed.

There were 14 Fish Engines, with the first appearing from Neilson & Co in November 1902, and the remaining five on their order in December the same year. There was a gap of just over a year before the next series began to arrive from Beyer Peacock in January 1904.

Originally 3,250-gallon tenders were paired with the B5s, but with the GCR introducing longer non-stop runs by the time the LNER became their owners 4,000-gallon tenders were attached. All the tenders had some minor detail variations, and during LNER days there were numerous re-allocations within the class, although the larger tenders were retained until withdrawal.

Class 8G (LNER Class B9) 4-6-0

Robinson's second goods type for the Great Central was produced in the autumn of 1906 by Beyer Peacock of Manchester, and was a follow-up order for the company. Earlier that same year, Beyer Peacock had completed the contract to build the 8F class, the famous Immingham 4-6-0s.

The new goods engines were seen as a smaller wheeled version of the B5, and totalled ten in all, carrying GCR running numbers 1105–1114.

The steam raising plant consisted of a 5 ft 0 in diameter boiler, as used on Robinson's Atlantics, and the class 8C and 8F designs. This boiler in saturated form housed 226 2 in diameter small tubes, and provided a heating surface of 1,818 sq ft. The firebox though, at 7 ft 9 in long, was shorter than fitted to other ex-GCR types with this boiler, and provided a grate area of only 23·75 sq ft. Until they became LNER locomotives, the small wheeled Fish Engines saw little change, although No 1112 had received a 4 ft 9 in diameter boiler from B5 4-6-0 No 182 in 1910, which was removed in 1919. The 4 ft 9 in diagram 17 boilers with superheaters began to appear on the B9s after the Grouping, and like their larger brethren were re-classified B9/2 until 1937. The diagram 17 boilers were pitched higher than the original design, and needed a number of detail alterations to comply with, among other restrictions, the LNER loading gauge.

The increased pitch of the superheated boilers enabled the firebox to clear the trailing coupled axle, while other detail changes included the removal of the Robinson pattern chimney, and its replacement on some locomotives by what was referred to as the 'flowerpot' design. Superheating on these locomotives, as on many other inherited types, saw the provision of Gresley's anti-

GCR Class 8G (LNER Class B9) 4-6-0 – Building and withdrawal

GCR No	Built	BP Works No	1924 No	1st 1946 No	2nd 1946 No	BR No	Withdrawn
1105	9/06	4806	6105	1342	1469	61469	4/49
1106	9/06	4807	6106	1343	1470	61470	11/48
1107	9/06	4808	6107	1344	1471	—	11/47
1108	9/06	4809	6108	1345	1472	—	6/47
1109	9/06	4810	6109	1346	1473	—	12/47
1110	9/06	4811	6110	1347	1474	—	10/47
1111	10/06	4812	6111	1348	1475	61475	5/49
1112	10/06	4813	6112	1349	1476	61476	8/48
1113	10/06	4814	6113	1350	1477	—	8/47
1114	10/06	4815	6114	1351	1478	—	12/47

Notes:
1. All locomotives were built by Beyer Peacock & Co.
2. No 6111 was withdrawn in July 1939 and re-instated for further service.

vacuum valve atop the smokebox behind the chimney, and the replacement of the Ramsbottom safety valves by the familiar Ross pop variety set to lift at 200 lb/sq in.

Two outside 19 in by 26 in cylinders were fitted, operated by Stephenson valve gear, with steam chest and slide valves inside the frames. Only No 6109 received piston valves with 21 in diameter cylinders. Despite the reported improvement in performance, the rest of the class was not similarly dealt with, perhaps a sound commercial decision on the grounds of cost and the nature of the work which befell the B9s.

As with other pre-Grouping types, such features as piston tail rods were removed and other detail changes made, including altered smokebox door fastening and the repositioning of lamp irons. The smokebox ash ejectors fitted to the class from new were retained by about half their number, but the equipment was repositioned on the superheated locomotives, with pipes on the right-hand side of the smokebox. Steam-operated brakes were in place in the locomotive, with shoes carried ahead of the wheels, and a vacuum ejector provided for the trains' vacuum brake system. The new small-wheeled 'Fish Engines' differed from other GCR types in having sanding applied to the front of the intermediate as well as the leading coupled wheels.

Beyer Peacock Order No 9456 covered the construction of the ten locomotives – six in September and four in October 1906. All were superheated between October 1924 and April 1929. They were renumbered by the LNER by the simple expedient of adding 5,000 to their previous numbers. Tenders paired with the class were 4,000 gallons capacity, carrying six tons of coal – the standard Robinson, GCR six-wheeled design. Coal rails plated over were used on No 6110, while the remainder had plain sheet steel fenders at the sides of the coal space. No 6110's original tender had been transferred to various J11 0-6-0s, while the tender it ran with in normal service had come from Class D9 4-4-0 No 6038.

Four of the class survived to nationalisation in 1948.

MIXED-TRAFFIC LOCOMOTIVES

While the majority of Robinson's 4-6-0 designs inherited by the LNER had outside cylinders, his first essay in what might be termed the mixed-traffic category were inside-cylinder machines. In fact, the Class 1A (LNER B8 Glenalmond) was essentially a smaller wheeled variant of the Sir Sam Fay class. The Glenalmonds appeared in 1913, when their predecessors had been at work for around a year. Only one locomotive, carrying the running number 4, represented the class for more than twelve months, as the second Glenalmond did not appear until 1914.

The new 4-6-0s outshopped by Gorton Works perpetuated some of the same design faults as their progenitors, and it has been suggested that the shortcomings of the Sir Sam Fay class must already have been known at the time the new type was ordered. Certainly, the order for 10 more in 1914, over

---

GCR Class 1A (LNER Class B8 Glenalmond) 4-6-0 – Building and withdrawal

GCR No	Built	1924 No	Name	1st 1946 No	2nd 1946 No	Withdrawn
4	6/13	5004	Glenalmond	1331	1349	11/47
439	7/14	5439	Sutton Nelthorpe	1332	1350	8/47
440	8/14	5440		1333	1351	10/47
441	9/14	5441		1334	1352	5/47
442	9/14	5442		1335	1353	3/49
443	10/14	5443		1336	1354	3/48
444	10/14	5444		1337	1355	9/48
445	11/14	5445		1338	1356	8/47
446	11/14	5446	Earl Roberts of Kandahar	1339	1357	4/49
279	12/14	5279	Earl Kitchener of Khartoum	1340	1358	8/48
280	1/15	5280		1341	1359	3/47

Notes:
1. All locomotives were built at Gorton
2. No Works Numbers carried
3. None carried its BR Number.

two years after the introduction of the GCR Class 1 (LNER Class B2), is curious. Although mixed-traffic locomotives, most of their work involved goods trains, with some excursion and stopping passenger duties.

The boiler fitted to No 4 *Glenalmond* in 1913 later became LNER diagram 13, following changes to the number of tubes, flues, and superheater elements. Again, the boiler originally fitted, with its 157 2¼ in small tubes and 24-element superheater proved inadequate, so the locomotives which began appearing from 1914 had 139 small tubes, and 28-element superheaters. As in the Sir Sam Fay class, the final boiler design used on the Glenalmonds housed only 116 small tubes, but was not installed until after the Grouping. The class was also the first GCR type to be fitted with the Robinson top feed arrangement, with the feed clacks mounted on the front boiler ring – this equipment was later removed by the LNER.

Some B8s were fitted with the draught retarding apparatus, to protect the superheater elements, while by the Grouping all had the Robinson steam circulating valves. The same firebox was installed on the new 4-6-0s as had been seen on the Sir Sam Fay class. Like their larger counterparts, Nos 279, 443 and 445 were fitted with oil-burning equipment in May and June 1921.

Below the footplate, the Glenalmonds were carried on 5 ft 7 in coupled wheels, and 3 ft 6 in wheels on the leading bogie. Their two 21½ in by 26 in inside cylinders were operated by Stephenson valve gear, with

10 in diameter piston valves. Here too the same faults demonstrated themselves, with poor steam circuit design, and restricted clearances. The total wheelbase of the mixed-traffic B8 was only 7 in shorter than the B2. As LNER locomotives in their final form they possessed a nominal tractive effort of 27,445 lb. It has been suggested that these two classes of inside-cylinder 4-6-0 were strongly influenced by the Caledonian Railway Cardean and 908 classes. By contemporary standards they were very large machines, and although British locomotive designers do seem to have had an abhorrence of showing the 'works' of steam locomotives, it was a surprising layout to adopt in both cases.

Four years after they became LNER property, Nos 5443 and 5004 had their cylinders lined-up to 20 in diameter, and through this the tractive effort came down to 23,750 lb. The first of these, No 5443 was fitted with the original size cylinders again in 1935.

Locomotive main suspension was underhung leaf springs as in the earlier designs, while the brakes and sanding arrangements followed previous patterns. A notable feature of the Glenalmond class, and one which distinguished the type from the B2 or Sir Sam Fay type, was the continuous single-level running boards, with the large coupled wheel splashers as before. Lubrication methods were altered a number of times on these locomotives, with the original Wakefield mechanical lubricators replaced during World War I by Robinson's Intensifore arrangement. During the mid-1930s this in its turn was replaced by mechanical lubrication with either Detroit or Eureka pattern equipment.

The B6 Class – Robinson's 5 ft 8 in 4-6-0s
The three locomotives of Class 8N (which became LNER Class B6 at Grouping in 1923) had wheels and motion interchangeable with the O5 class 2-8-0. The first appeared in 1918, and remained the solitary example for some three years; it appears to have been constructed as something of an experiment. The company seemed to be planning comparative trials with a smaller-wheeled version of the Lord Faringdon class, which had been ordered along with the 2-8-0s from Gorton Works. However, the trials did not materialise, and the decision to

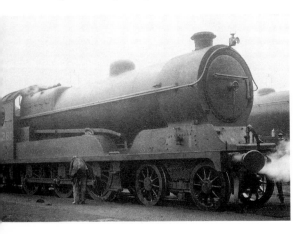

One of the inside-cylinder Glenalmond 4-6-0s, No 5280 in LNER colours, its large boiler giving a foreshortened appearance in view of the fairly short wheelbase of this type. *Lens of Sutton*

GCR Class 8N (LNER Class B6) 4-6-0. Building and withdrawal

GCR No	Built	1924 No	1st 1946 No	2nd 1946 No	Withdrawn
416	7/18	5416	1328	1346	11/47
52	3/21	5052	1329	1347	12/47
53	4/21	5053	1330	1348	12/47

Notes: All built at Gorton. No Works Numbers issued

build further locomotives of a mixed-traffic type resulted in more orders for the 9Q (LNER Class B7) and the ex-GCR Class 8N (LNER Class B6) totalled just three locomotives. The remaining two Class 8N were built at Gorton in 1921, and carried running numbers 52 and 53.

The preceeding mixed-traffic design, the Glenalmond, suffered from a number of design faults, including inadequately proportioned axleboxes, due to the need to accommodate the large cylinders between the frames. Placing the cylinders outside the frames with a diameter only slightly smaller than previously at 21 in allowed more space between the frames. While the boiler diameter was the same as the 1A (LNER B8), it was more than 2 ft shorter at 15 ft 0 in, or 15 ft 4³/₈ in between tubeplates. The firebox provided 11 sq ft more heating surface than the 1A or the succeeding four-cylinder Class 9Q 4-6-0 (LNER Class B9) and although the length was the same, it was deeper at front and back ends by almost 12 in. Although the boilers were shorter, with a heating surface of 2,123 sq ft compared with the 2,387 sq ft of their inside-cylinder predecessors, they proved to be a more free-steaming design.

With the arrangement of outside cylinders in the new Class 8N, it was possible to provide larger driving axleboxes, without the attendant problems of the inside-cylinder designs. Modifications in LNER days were few, with the 10 in piston valves fitted with new heads and narrow rings in the mid-1930s, though unlike other GCR types the B6 4-6-0s retained their piston tail rods until withdrawal.

Other changes in LNER days were minor, and largely concerned the diagram 15B boiler, from which the Robinson top feeds disappeared, and where the Gresley pattern anti-vacuum valves were fitted. Despite minor changes to other boiler fittings, various patterns of dome cover and chimney, these locomotives were never altered to comply fully with the LNER loading gauge.

When the last two of the class appeared in 1921, the Great Central had adopted a cab with side windows, instead of the large cutaway in the side sheets. Nos 5052 and 5053 had still larger driving axleboxes than No 5416. None received the once standard ash ejector. Lubrication saw some change, when Robinson's original Intensifore system was replaced by Wakefield mechanical lubricators in the mid-1930s, feeing oil to the coupled axle journals, in addition to steam chests and cylinders.

Tenders paired with these locomotives remained essentially unaltered throughout the locomotives' 30-year lifespan. These were standard 4,000 gallons water capacity, fitted with water pick-up gear, and carrying six tons of coal. The first of the class differed slightly in having a tender fitted with a patent form of outside bearing axlebox – the Iracier type.

Class B7 4-6-0 – The 'Black Pigs'

The final design of 4-6-0 from the Robinson stable was a 5 ft 8 in mixed-traffic type derived from the Lord Faringdon class. According to contemporary reports, the GCR Class 9Q was built for working vacuum-fitted fish trains between Marylebone and Grimsby. They were most attractive locomotives, coming from Gorton Works, Vulcan Foundry, and Beyer Peacock & Co, over a three-year period between 1921 and 1924.

With the new design, four 16 in diameter cylinders were placed in line under the smokebox, with the inner pair driving onto the leading axle, and the outer pair the middle axle. With a tractive effort of 29,500 lb, the company was confident that its new four-cylinder 4-6-0s would accomplish all that was expected of them. In service their nickname was acquired in company with such others as 'Colliers' Friends', due to their propensity to consume large quantities of fuel. This reputation was not entirely justified, since in comparison with other types on similar duties for the LNER, the former GCR locomotives actually burned less coal. A total of 28 was built as Class 9Q before 1923, with ten more emerging from Gorton Works as LNER Class B7 in 1923/24.

The boilers on the new 4-6-0s were identical with those on the Sir Sam Fay and Glenalmond classes, although the smaller coupled wheels enabled improvements to be made at the firebox end. The ten locomotives built by the LNER had reduced height boiler mountings to conform with the new loading gauge. The boiler was the final development of the diagram 13 type, with 116 small tubes, and a 28-element superheater, also used on the B3 4-6-0, with the Robinson pattern cast iron header, and steam circulating valve controlled by operation of the blower. In original form, the B7s were fitted with top feed, and on the centre boiler ring, the dome housed a double-seated balanced regulator valve.

The waisted-in Belpaire topped firebox was surmounted by pop safety valves set to lift at 180 lb/sq in pressure, though the two outside contractors seemed to show 200 lb/sq in pressure. The outside length of the box was 8 ft 6 in, and with the smaller coupled wheels it was possible to increase ashpan dimensions. At the front of the box, the height was 6 ft $6^{1}/_{8}$ in, while at the rear it was 5 ft $0^{5}/_{8}$ in, with a grate area of 26 sq ft. The extra

GCR Class 9Q (LNER Class B7) 4-6-0 Building and withdrawal

GCR No	Built	Builder	Works No	1924 No	1946 No	1st BR No	2nd BR No	Date	Withdrawn
72	5/21	Gorton Works	—	5072	1360	—	—	—	9/48
73	6/21	Gorton Works	—	5073	1361	—	—	—	3/49
78	7/21	Gorton Works	—	5078	1362	—	—	—	4/49
36	9/21	Vulcan Foundry	3478	5036	1363	—	—	—	6/48
37	10/21	Vulcan Foundry	3479	5037	1364	—	—	—	6/48
38	10/21	Vulcan Foundry	3480	5038	1365	—	61702	5/49	6/49
458	10/21	Vulcan Foundry	3481	5458	1366	—	—	—	12/48
459	10/21	Vulcan Foundry	3482	5459	1367	—	61703	5/49	9/49
460	10/21	Vulcan Foundry	3483	5460	1368	—	—	—	10/48
461	11/21	Vulcan Foundry	3484	5461	1369	—	—	—	8/48
462	11/21	Vulcan Foundry	3485	5462	1370	—	—	—	11/48
463	11/21	Vulcan Foundry	3486	5463	1371	—	—	—	1/49
464	11/21	Vulcan Foundry	3487	5464	1372	—	—	—	9/48
465	8/21	Gorton Works	—	5465	1373	—	—	—	8/48
466	10/21	Gorton Works	—	5466	1374	—	—	—	9/48
467	2/22	Gorton Works	—	5467	1375	—	61704	4/49	6/49
468	3/22	Gorton Works	—	5468	1376	—	—	—	12/48
469	4/22	Gorton Works	—	5469	1377	—	61705	5/49	2/50
470	5/22	Gorton Works	—	5470	1378	—	—	—	8/48
471	6/22	Gorton Works	—	5471	1379	—	—	—	2/49
472	6/22	Gorton Works	—	5472	1380	—	—	—	8/48
473	7/22	Gorton Works	—	5473	1381	—	61706	4/49	12/49
474	8/22	Gorton Works	—	5474	1382	—	61707	4/49	6/49
31	7/22	Beyer Peacock & Co	6107	5031	1383	—	—	—	5/48
32	7/22	Beyer Peacock & Co	6108	5032	1384	—	—	—	8/48
33	8/22	Beyer Peacock & Co	6109	5033	1385	—	—	—	1/49
34	8/22	Beyer Peacock & Co	6110	5034	1386	—	(61708)	—	6/49
35	8/22	Beyer Peacock & Co	6111	5035	1387	—	61709	5/49	1/50
475	8/23	Gorton Works	—	5475	1388	—	61710	5/49	2/50
476	8/23	Gorton Works	—	5476	1389	—	—	—	2/49
477	9/23	Gorton Works	—	5477	1390	—	—	—	11/48
478	10/23	Gorton Works	—	5478	1391	61391	61711	4/49	7/50
479	11/23	Gorton Works	—	5479	1392	—	61712	5/49	6/49
480	11/23	Gorton Works	—	5480	1393	—	—	—	8/48
481	12/23	Gorton Works	—	5481	1394	—	—	—	4/48
482	12/23	Gorton Works	—	5482	1395	—	—	—	11/48
—	2/24	Gorton Works	—	5483	1396	61396	61713	4/49	9/49
—	3/24	Gorton Works	—	5484	1397	—	—	—	6/48

Notes:
1. Nos 5483/4 did not receive GCR numbers
2. Only two locomotives, Nos 1391/6 had 1st BR numbers, created by adding 60,000 to the 1946 numbers
3. Locomotives not listed under '1st BR No' went direct from the '1946 No' to '2nd BR No' in April/May 1949
4. No 1386 was allocated No 61708, but did not receive it.

space available over the coupled wheels, together with the positioning of the trailing axle 7 in further forward than on the B3, enabled the fitting of a rear ashpan damper door.

Although the B7s retained their diagram 13 boilers to the end, in later LNER days Edward Thompson had planned to reboiler the class with the new B1 boiler, and provide two outside cylinders only. Some preparatory work was done, and diagrams showed a tractive effort of 28,600 lb for the modified design, with a boiler working pressure of 220 lb/sq in, and a reduced overall length for the type, with weight down by $5\frac{1}{2}$ tons. However, the idea never got beyond the drawing board.

Construction
The four cylinders were provided with 8 in diameter piston valves. As in the earlier four-cylinder designs, the inside pair was arranged for inside admission, and the outside pair for outside admission. The valve rods of the outside pair were operated by rocking shafts, while the still short travel of $4\frac{3}{4}$ in was employed, using two sets of Stephenson valve gear. The valves were provided with another Robinson innovation, release rings which acted as drifting valves, to release excessive water or steam pressure into the steamchest. The crank axles were standard with the 'Pom Pom' 0-6-0s, while the four locomotives in the last batch, Nos 5480/2/3/4, had larger steamchests, and slightly altered external appearance. Other changes included vertical wrapper plates, and snifting valves in the front end covers. Later, a number from the pre-1923 builds received these improved cylinders, resulting in a better layout of the steam circuit. New front ends, with alterations to the front portion of the mainframes, were the outcome of incidents of frame distortion near the cylinders, which occurred in the 1940s.

Outside the frames, connecting and coupling rods were deeply fluted, with the heavy cast motion bracket supporting the double slidebars, and incorporating a footstep, as in the earlier four-cylinder design. The 18-spoke 5 ft 8 in coupled wheels were provided with case-hardened wrought-iron axleboxes, having gunmetal bearings faced with white metal. The leading axle journals were 8 in diameter but only 9 in long, due to the proximity of the crank axle webs, while the remainder were 12 in long. Spring-supported oil pads were once again employed to reduce maintenance time, while lubrication of the cylinders, steamchests, and four driving axleboxes used the Intensifore system.

The leading bogie with its 10-spoke 3 ft 6 in diameter wheels followed standard GCR construction practice, with both main suspension and side control effected by means of laminated springs. Running boards were raised over the cylinders, with the same large coupled wheel splasher as in earlier designs. The cab included two sliding windows in the sides, and an extension to the roof. While the majority of the class were paired with a standard GCR 4,000-gallon tender with water pick-up gear and carrying six to seven tons of coal, the first locomotive No 72 (LNER No 5072) had been equipped with Unolco oil-firing apparatus. The installation, which included a large oil tank mounted in the tender coal space, was removed only three months after the locomotive was completed.

The GCR-built locomotives came out of Gorton Works at regular intervals between May 1921 and October 1921, then February to August 1922. Meanwhile Beyer Peacock and Vulcan Foundry had built batches of ten and five respectively, between September 1921 and August 1922. The LNER-built locomotives emerged from Gorton Works between August 1923 and March 1924.

During their LNER service, most changes centred on the adoption of that company's standard practices, hence the disappearance of top feeds and superheated header discharge valves. Gresley anti-vacuum valves appeared behind the chimney, which also saw a number of minor variations. Four diagrams were issued by the LNER covering several of these changes, ranging from removal of top feed, to classifying locomotives over 13 ft 0 in high as B7/1, and those less than 13 ft 0 in as B7/2. The size of the B7 cab was not reduced to conform to the new loading gauge, although the entire class eventually received reduced height boiler mountings.

The GCR types in operation
Robinson's passenger types, making up LNER classes B1 (later B18), B2 (later B19), B3 and B4, amounted to 24 locomotives, and all suffered from some design weaknesses. In

fact, the shortcomings were common to almost all the 4-6-0 designs – his persistence with short-travel valves, and his curious inability to provide large enough firegrates, with generously proportioned ashpans.

The former Great Central passenger types did not, generally speaking, appear to excel themselves under normal working conditions, on their varied passenger duties. The two B1s had moved from Sheffield to Immingham by the Grouping, going south to Neasden and Woodford until the early days of World War II. Mainly used on secondary turns, they transferred between various East Midlands sheds in the 1940s, before being withdrawn in December 1947. The notorious B2 class, whose shortcomings were discovered soon after they were put into traffic from Gorton, had been relegated to fast, fitted goods trains between Manchester and Marylebone by the end of World War I. On London trains, the B2s were easily outclassed by the Director series of 4-4-0s, and the later four-cylinder Lord Faringdon class 4-6-0. Sheffield and Immingham were popular locations for B2s in the late 1920s and 1930s, covering a wide variety of duties, including Bradford to Marylebone trains. An eventful incident occurred to No 5423 *Sir Sam Fay* when working the 10.00 am express from Marylebone in January 1933, during which journey it collided with a Class J39 0-6-0 on a train of empty wagons at Loughborough.

As a result of that accident, involving a lightweight train of only five coaches, or around 175 tons, a series of brake trials was conducted. Involved in the trials was a steam brake fitted B2, and one of the ex-GNR Class C1 Atlantics, with vacuum brake equipment. With trains of five and eight coaches, the Atlantic was reckoned to perform better than the steam-braked B2s. Performance-wise the Atlantic was consistently superior to No 5427, although the latter managed a maximum of 84 mph during the tests.

In the late 1930s and early 1940s B2s based at Lincoln had been working the boat trains from Lincoln to York, but with World War II vastly increased loads became commonplace. The 1940s saw more re-allocations and extended workings for other classes, with B3s going to Woodford, and replaced at Immingham by B2s. London trains to and from the East Midlands were heavy in the 1940s,

although generally workings with B2s were of modest weight, and they were reportedly speedy locomotives.

The six locomotives of the B3 Lord Faringdon class were divided between Gorton and Immingham, although the latter shed had little express passenger work. Gorton locomotives were sent out on the slower London workings. In early post-Grouping days they found their way onto the former Great Northern main line for a short time, and although they were not so successful as the ex-GNR Atlantics overall, they would perhaps have performed better on trains demanding smart acceleration. While GCR 4-6-0s were put to work on the LNER Pullman workings, former GNR locomotives were preferred, and in general the GCR 4-6-0s were relegated to other work. The Immingham class locomotives began their careers at Neasden, Gorton, and Grimsby, working express goods and fish traffic, and express passenger work in the London area. The LNER replaced them on passenger work at Sheffield by Ivatt's large Atlantics, and into the 1930s they could be found in the West Yorkshire area. From trials carried out in the 1930s involving No 1097 *Immingham*, Gresley's highly successful Class V2 2-6-2 eventually emerged. In the 1930s re-allocation of B4s to East Anglia took place and the distribution of the class remained largely unaltered up to nationalisation. The Imminghams did not survive long after nationalisation, with No 1097 the last to go in 1950.

Another interesting series of trials involving former GCR 4-6-0s was held by the LNER in 1927, to determine the needs for electric traction over the Woodhead route through the Pennines. The LNER trials involved B3 No 6164 *Earl Beatty*, which by all accounts put in some brisk performances, and fast point to point times. The B3 recorded a maximum power output of 1,010 equivalent drawbar hp at Dunford Bridge, on one of the test runs.

The freight and mixed-traffic 4-6-0s of Great Central origin were stabled at Grimsby, Neasden, Gorton, Woodford, Immingham, and Lincoln. The B5s or 'Fish Engines' were primarily used on fish trains throughout their lives, although towards the end they were rostered to work over unfamiliar routes of the former Cheshire Lines

Committee, having been transferred to Trafford Park (Manchester). The three Class B6 locomotives in LNER service worked throughout the system on goods and passenger services, with the latter mainly excursion trains.

The B7 'Black Pigs' with their reputations as coal eaters, were distributed to various locations on passenger and freight work, often deputising for B3s. In the 1940s, the arrival of the new Thompson B1 4-6-0s was causing some displacement, despite which a number survived to be handed over to British Railways.

The remaining two classes, B8 and B9, representing freight and mixed-traffic types respectively, also disappeared early in BR days. The B9s, having originally been stabled at Lincoln and Gorton saw a number of moves before finally ending up on all types of work over ex-Cheshire Lines metals in the late 1940s.

The 11 Glenalmonds were a mixed-traffic type, and originally handled fast goods and fish trains, but were quickly replaced on these duties by the B7s. During the 1930s, B8s were employed for a time on an accelerated coal train working betwen Anneseley and Woodford, also being used on a great deal of excursion work in the Nottingham area. During World War II, in common with other GCR designs, heavy goods trains and troop traffic were the lot of this class, taking some locomotives far from their home territory.

On the whole, the LNER throughout the 1920s and 1930s persevered with 4-6-0 designs which on other lines would in all probability have been scrapped. The GCR types had a number of inherent design faults which were not easily or economically remedied, and the Robinson designs were severely criticised for their fuel consumption. Performances could be quite good and, as a number of recorded trial runs bears witness, when put to the test they were capable and speedy machines. They were in a number of cases complicated locomotives with a variety of gadgets included in their construction, which had no place in steam locomotive design, especially in post-World War II years.

Former Great Central Railway 4-6-0s Types – Leading Dimensions

Class	8C (LNER B1 – later B18)	1 (LNER B2 – later B19)	9P (LNER B3)
Coupled wheel diameter	6 ft 9 in	6 ft 9 in	6 ft 9 in
Bogie wheel diameter	4 ft 4 in	3 ft 6 in	3 ft 6 in
Wheelbase	26 ft 9$^{1}/_2$ in	28 ft 10 in	28 ft 10 in
Boiler – length	15 ft 0 in	17 ft 3 in	17 ft 3 in
– diameter	5 ft 0 in	5 ft 6 in	5 ft 6 in
– tubes: small	226 × 2 in	139 × 2$\frac{1}{4}$ in/116 × 2$\frac{1}{4}$ in	138 × 2$\frac{1}{4}$ in/116 × 2$\frac{1}{4}$ in
– tubes: large	—	24 × 5$\frac{1}{4}$ in/28 × 5$\frac{1}{4}$ in	24 × 5$\frac{1}{4}$ in/28 × 5$\frac{1}{4}$ in
Heating surface – tubes	1818.00 sq ft	2020.00 sq ft/1881.00 sq ft	2020.00 sq ft/1881.00 sq ft
– firebox	133.00 sq ft	163.00 sq ft/ 163.00 sq ft	163.00 sq ft/ 163.00 sq ft
– Total	1951.00 sq ft	2183.00 sq ft/2044.00 sq ft	2183.00 sq ft/2044.00 sq ft
Superheater – elements	—	24 1$\frac{1}{16}$ in/28 1$\frac{1}{16}$ in	24 1$\frac{1}{16}$ in/28 1$\frac{1}{16}$ in
– heating surface	—	294.00 sq ft/343.00 sq ft	294.00 sq ft/343.00 sq ft
Firebox – length	8 ft 6 in	8 ft 6 in	8ft 6 in
– grate area	26.24 sq ft	26 sq ft	26 sq ft
Cylinders	21 in × 26 in/19 in × 26 in	20 × 26/21$\frac{1}{2}$ × 26 in	16 × 26 in
Tractive effort	21,685 lb/17,729 lb	19,644 lb/22,700 lb	25,145 lb
Fuel capacity – coal	6 tons	6 tons	6 tons
– water	4,000 gallons	4,000 gallons	4,000 gallons
Weight in working order	*Tons cwt*	*Tons cwt*	*Tons cwt*
– locomotive	72 18	71 00	75 04	79 02
– tender	48 06	48 06	48 06	48 06
– Total	121 04	119 06	123 10	127 08

Class	8F (LNER B4)	8 (LNER B5)	8N (LNER B6)
Coupled wheel diameter	6 ft 7 in	6 ft 1 in	5 ft 8 in
Bogie wheel diameter	3 ft 6 in	3 ft 6 in	3 ft 6 in
Wheelbase	26 ft $9\frac{1}{2}$ in	26 ft $1\frac{1}{2}$ in	27 ft 6 in
Boiler – length	15 ft 0 in	15 ft 0 in	15 0 in
– diameter	5 ft 0 in	4 ft 9 in	5 ft 6 in
– tubes: small	226 × 2 in	207 × 2 in	116 × $2\frac{1}{4}$ in
– tubes: large	—	—	28 × $5\frac{1}{4}$ in
Heating surface – tubes	1818.00 sq ft	1665.00 sq ft	1641.00 sq ft
– firebox	133.00 sq ft	130.00 sq ft	174.00 sq ft
– Total	1951.00 sq ft	1795.00 sq ft	1815.00 sq ft
Superheater – elements	—	—	28 × $1\frac{1}{16}$ in
– heating surface	—	—	308.00 sq ft
Firebox – length	8 ft 6 in	7 ft 9 in	8 ft 6 in
– grate area	26.24 sq ft	23.5 sq ft	26.24 sq ft
Cylinders	19 in × 26 in	19 in × 26 in	21 in × 26 in
Tractive effort	18,178 lb	19,672 lb	25,798 lb
Fuel capacity – coal	6 tons	6 tons	6 tons
– water	4,000 gallons	4,000/3,250 gallons	4,000 gallons
Weight in working order	*Tons cwt*	*Tons cwt* / *Tons cwt*	*Tons cwt*
– locomotive	70 14	65 02 / 65 02	72 18
– tender	48 06	48 06 / 44 03	48 06
– Total	119 00	113 08 / 109 05	121 04

Class	9Q (LNER B7)	1A (LNER B8)	8G (LNER B9)
Coupled wheel diameter	5 ft 8 in	5 ft 7 in	5 ft 4 in
Bogie wheel diameter	3 ft 6 in	3 ft 6 in	3 ft 6 in
Wheelbase	28 ft 3 in	28 ft 3 in	26 ft $1\frac{1}{2}$ in
Boiler – length	17 ft 3 in	17 ft 3 in	15 ft 0 in
– diameter	5 ft 6 in	5 ft 6 in	5 ft 0 in
– tubes: small	116 × $2\frac{1}{4}$ in	139 × $2\frac{1}{4}$ in/116 × $2\frac{1}{4}$ in	226 × 2 in
large	28 × $5\frac{1}{4}$ in	24 × $5\frac{1}{4}$ in/28 × $5\frac{1}{4}$ in	—
Heating surface – tubes	1881.00 sq ft	2020.00 sq ft/1881.00 sq ft	1818.00 sq ft
– firebox	163.00 sq ft	163.00 sq ft/ 163.00 sq ft	133.00 sq ft
– Total	2044.00 sq ft	2183.00 sq ft/2044.00 sq ft	1951.00 sq ft
Superheater – elements	28 × $1\frac{1}{16}$ in	24 × $1\frac{1}{16}$ in/28 × $1\frac{1}{16}$ in	—
– heating surface	343.00 sq ft	294.00 sq ft/343.00 sq ft	—
Firebox – length	8 ft 6 in	8 ft 6 in	7 ft 9 in
– grate area	26 sq ft	26 sq ft	23.75 sq ft
Cylinders	16 in × 26 in (4)	$21\frac{1}{2}$ in × 26 in	19 in × 26 in
Tractive effort	29,952 lb	27,445 lb	22,438 lb
Fuel capacity – coal	6 tons	6 tons	6 tons
– water	4,000 gallons	4,000 gallons	4,000 gallons
Weight in working order	*Tons cwt*	*Tons cwt*	*Tons cwt*
– locomotive	79 10	74 07	67 06
– tender	48 06	48 06	48 06
– Total	127 16	122 13	115 12

CHAPTER FOUR

B12 – A SUCCESSFUL GER CONTRIBUTION

In common with many other British railways during the early years of this century, the Great Eastern Railway saw its traffic increasing sufficiently to need a more powerful passenger locomotive. Following the examples set by its contemporaries, the GER chose the 4-6-0 design, and under the guidance of F. V. Russell, in charge of locomotive design at Stratford Works, the new S69 class locomotives were produced from 1911. In its original form it was a very attractive design, as were a number of other types built during S. D. Holden's reign as locomotive superintendent. In later years the design was the subject of numerous alterations, particularly during the Gresley era, when a further 10 locomotives were ordered from Beyer Peacock. It is interesting to speculate that these might not have materialised had it not been for the accident which befell the Southern Railway River class 2-6-4T at Sevenoaks in 1927. The LNER had planned to produce a tank engine design with this wheel arrangement, but the criticism of such layouts following the Sevenoaks incident influenced the abandonment of the LNER scheme, and resulted in the ordering of more of the B12 type, as it became in the post-Grouping classification.

The many variations on the B12 theme included poppet valve gear, feedwater heaters, alterations to the platework, and other detail changes. They were among the smallest 4-6-0s being built at that time, but were intended for use on some of the Great Eastern's more exacting main line passenger schedules. The S69 class – otherwise the 1500 class – had two inside cylinders, 6ft 6in coupled wheels, and were lightweight machines with a low axle loading recorded by the LNER as 15 tons 13 cwt maximum, but was stated to be originally 16 tons, 14 tons, and 14 tons, on the driving, centre and trailing coupled axles. This light weight gave the class a wide route availability, a feature used

to some extent in later years by the LNER, when the class was at work in Scotland.

The overall wheelbase was kept to 48 ft 3 in with the standard short GER tender, which allowed the locomotives access to the many 50ft diameter turntables on the GER system. A drawback to the design of the footplate, with a distance between tender and firehole of some 8 ft, meant that the fireman had to use a very long-handled shovel when firing.

Not all the locomotives were built at Stratford, and out of the 70 which were constructed before the Grouping 20 came from Beardmore & Co of Glasgow in 1920/21. A final batch of 10 was ordered from Beyer Peacock in Manchester by the LNER in 1928. While No 1506 had a very short career – only six months as a result of an accident at Colchester in July 1913, after which it was scrapped – the majority of the class survived for many years. Out of a total of eighty B12s, excluding the written-off No 1506, only eight did not survive until nationalisation, and the greater part of the class continued to see regular service into the mid-1950s. Withdrawal took place quite quickly, with the

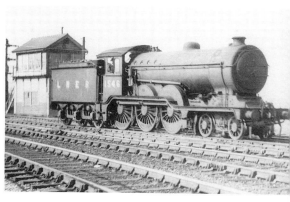

Gresley's large boilered rebuild of the GER 4-6-0 as Class B12, although minus the decorative valance, is still an attractive design. *Lens of Sutton*

spread of diesel traction in the Eastern Region, the last going in 1961.

From new, the 1500 class was a successful design, allocated initially to Norwich services, later including the Great Eastern's principal expresses to and from London. Changes occurred after their takeover by the

---

GER S69/1500 (LNER B12) 4-6-0 – Leading details

Class	LNER B12
Previous owner/class	GER S69/1500
Built	12/1911–10/1928
Withdrawn	6/1945–9/1961
Running numbers	
– original	1500–1570, 8571–8580
– 1924	8500–8580
– 1942	7415–7494
– 1946	1500–1580
– British Railways	61500–61580

Wheel diameter	
– coupled	6 ft 6 in
– bogie	3 ft 3 in
Wheelbase	28 ft 6 in

Length over buffers	57 ft 7 in

Axle load	Tons Cwt	Tons Cwt	Tons Cwt
– coupled	16 00	14 00	14 00
	Tons Cwt		
– bogie	20 00		

Boiler	
– diameter	5 ft $1\frac{1}{8}$ in
– length	12 ft 10 in between tubeplates
– tubes: small	191 × $1\frac{3}{4}$ in o/d
– tubes: large	21 × $5\frac{1}{4}$ in o/d

Heating surface	
– tubes	1123.00 sq ft
– firebox	143.50 sq ft
– Total evaporative	1266.50 sq ft

Superheater elements	21 × $1\frac{3}{8}$ in o/d
Superheater	286.40 sq ft

Working pressure	180 lb/sq in

Grate area	26.5 sq ft

Cylinders	
– number	2
– size	20 in × 28 in

Tractive effort	21,969 lb

Fuel capacities	
– coal	4 tons
– water	3,700 gallons

Weights (full)	Tons	Cwt
– locomotive	64	00
– tender	39	05
– Total	103	05

LNER in 1923 and British Railways in 1948, and while some experimental work was carried out they remained successful passenger 4-6-0s throughout their working lives.

Boiler Construction

The original boilers fitted to the new GER 4-6-0s were constructed in two rings, housing 191 tubes and 21 superheater flues. The barrel proper was 12 ft 6 in long, and 12 ft 10 in between tubeplates. The steam dome, carried on the rear boiler ring was 8 ft $3\frac{3}{4}$ in from the smokebox tubeplate, and housed the vertical grid type regulator. The superheaters, with twenty-one $1\frac{3}{8}$ in diameter elements housed in $5\frac{1}{4}$ in flue tubes, installed in locomotives Nos 1500–1512 and Nos 1515–1535, were of the Schmidt type. Superheater heating surface was reduced from 286.4 to 201.6 sq ft, when elements with short return loops were installed just before the Grouping. In Nos 1513 and 1514 Robinson type superheaters were fitted on an experimental basis, and used in No 1536 onwards from new, later becoming the standard type for the whole class. To protect the superheater elements from burning on locomotives fitted with the Schmidt pattern superheaters with long return loops, dampers were provided with the control mechanism carried on the left side of the smokebox.

By the time the LNER became owners there were already some detail variations between individual locomotives. The boilers were classed as diagram 25, including locomotives which had both patterns of superheater, a smaller number of tubes, reduced to 187 $1\frac{3}{4}$ in diameter. In the locomotives which sported an oval firehole door, only 165 tubes were fitted. These were originally known as diagram 99 boilers, but in 1938 the LNER chose to include them with diagram 25, most commonly fitted to Class B12.

Even before the LNER took over the type, examples of experimental work carried out included the installation of Stone's variable blastpipe on Nos 1505–1540 from new, and on Nos 1500–1504. Between 1925 and 1929 the LNER replaced these fittings – which purported to give the driver some degree of control over the opening of the blast nozzle – with the more conventional plain, circular type. Similarly, smokebox ash ejectors were fitted to Nos 1541–1570 from new, but removed by the LNER in the mid-1920s.

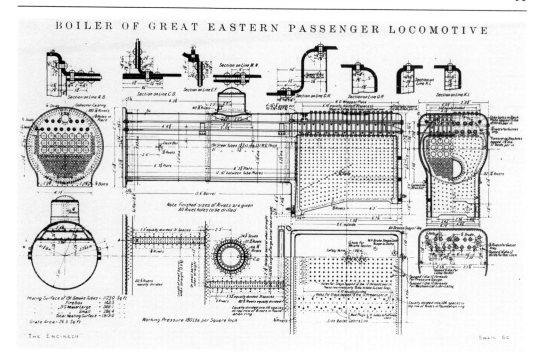

BOILER OF GREAT EASTERN PASSENGER LOCOMOTIVE

The firebox assembly consisted of a waisted-in Belpaire topped firebox, its steel outer casing stayed to a copper inner box, with an overall width of 4 ft 1½ in, and a length of 8 ft 6 in. The grate, with an area of 26.5 sq ft, had a flat (4 ft 3 in) rear portion over the trailing coupled axle, sloping down to the throatplate. On the firebox crown, four mounting points were installed for the Ramsbottom safety valves, although these too were changed to the pop type from the late 1920s.

Boiler feed was originally through a pair of live steam, non-lifting injectors, feeding into the side of the boiler through clack valves on the front boiler ring. The Davies & Metcalfe exhaust steam injector was tried from 1918 on Nos 1510/2/3/5/36, when one was fitted in place of a live steam injector on the left-hand side behind the cab footsteps. From No 1541 onward this was standardised, with the exhaust steam injector carried at the forward end of the locomotive, below the running boards. The later LNER series, Nos 8571–8580, also had this arrangement, despite the obvious difficulties in watching the overflow in the dark, since that pipe discharged at the front footstep. This method of boiler feed and the positioning of the injector was later discontinued when the locomotives were being rebuilt to Class B12/3 under Gresley.

Frames, Wheels and Motion

Traditional steel plate main frames were used, with the two inside 20 in by 28 in cylinders in a single casting acting as a frame stretcher at the front end. Stephenson valve gear was used, with the drive taken from the leading axle. The B12s with a weight of 20 tons on the leading bogie were heavy at the front; initially, due to the crank settings on the first five locomotives, some problems were encountered with axlebox wear.

To overcome these early problems, Nos 1500–1504, which had originally been equipped with forged driving crank axles and with cranks set at 180° out of phase with the adjacent wheel crank, were provided with built-up crank axles. In addition, the cranks were set to be in phase with the adjacent wheel crankpin, and heavier balance weights fitted. While the main coupled axleboxes were conventional plain bearing type, the rear bearing on No 1504 was given 1½ in side play, allowing more flexibility on curves, in a similar arrangement to that used on the 'Decapod' 0-10-0. The layout, in which the trailing axleboxes were self-centring, was fairly complex, and the trailing coupling rod was provided with a joint which allowed the rod to turn on its vertical axis to take account of the additional movement in the trailing axleboxes. This arrangement was

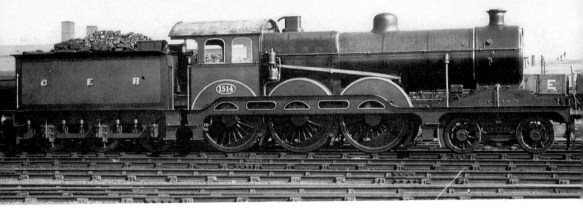

Here a broadside view of No 1514, in original livery as Class S69 is seen at Stratford in 1914. *L&GRP*

standard for all the 4-6-0s built by the Great Eastern, and the first four locomotives which were not so fitted originally were converted to this layout. The LNER-built series, Class B12/2, did not have this arrangement.

The coupled wheels were 6 ft 6 in diameter with 20 spokes. Main suspension was by means of underhung leaf springs, with inverted leaf springs and compensating beams forming the suspension on the leading bogie, with its 3 ft 3 in diameter 10-spoke wheels.

Brakes were air-operated, with shoes carried ahead of the wheels on single hangers, and the Westinghouse pumps mounted on the driver's side of the firebox. The brake system air reservoir was used also to operate the sanding equipment and water pick-up gear. A dispute with Westinghouse, which objected to the brake air reservoir being used for anything except its intended purpose, was overcome by segregating the reservoir into two components, so that feeds for auxiliary apparatus would not impinge on the integrity of the brake system reservoir.

The 20 in by 28 in cylinders were cast in one piece, integral with the 10 in diameter piston valves above, measuring 4 ft 1½ in over the bolting faces. The pistons carried three rings, with extensions on the tail rods. The two 10 in diameter valves, carrying four rings each, were fitted to two 1 in diameter valve rods with a square key and secured by a nut. The valve rods were operated by rocking shafts, imparting a maximum travel to the valves of only 4³/₁₆ in. After the 1923 Grouping, Gresley selected the class for a number of experimental alterations, which included Lentz poppet valve gear, and longer travel valves.

Above the footplate, and in general external appearance, the new locomotives were similar to the Claud Hamilton class 4-4-0s. Like the former NER types, the large roomy cab with two side windows was not typical of contemporary British practice. Although the sides and spectacle plate were sheet metal, the roof was constructed from wood, standard GER practice. This feature was extended to the locomotives built by Beyer Peacock in 1928, with minor alterations. These older style roofs were gradually replaced in the 1930s with a conventional steel design, although at least two locomotives were withdrawn still possessing the original wooden variety.

Originally, the running plate valances were perforated for decorative effect, and in their GER blue livery ornamental brass beading was applied to wheel splashers, the smokebox-boiler joint, and valances. Oval cast brass numberplates were fitted to the cab sides, and the securing ring around the smokebox door was bright finished. Locomotives built by the Great Eastern had a chimney topped with brass, but when the Beyer Peacock locomotives arrived, the chimney was a simple cast-iron affair. These new locomotives did not have the decorative valance of the earlier series. The cab sides, which initially had no beading where the arc of the coupled wheel splasher would have been, had this feature added at a later date, and the numberplate was moved lower down, inside it.

The majority of the B12s had the smaller Westinghouse air pumps of 6 in/6½ in diameter, on the right-hand side of the locomotive, but locomotives dispatched by the LNER to Scotland in 1931 had larger 8 in/8½ in diameter pumps, as did the rebuilt B12/3 series. Some of the latter did carry the small pumps, but the 8 in/8½ in variety was officially standard for the whole class. Later rebuilds to class B12/4 had the pumps moved 2 ft further forward.

When A. J. Hill was in office on the GER, snifting or anti-vacuum valves appeared for the first time on No 1519, carried on the smokebox behind the chimney, along with an external blower pipe on the right-hand side of the locomotive. All locomotives built new from 1913 had this fitting, and it was later applied to the rest of the class. One detail feature of the Great Eastern 4-6-0s which disappeared later was the bolting of guard irons to the front buffer beam. In the 1930s when some of the class were sent to Scotland for working over the former Great North of Scotland line these guard irons were removed and new irons substituted, which bolted onto the mainframes just behind the buffer beam.

Construction

As Class 1500, or S69, the new 4-6-0s produced under Holden consisted of 71 locomotives, including the one which due to the accident at Colchester in 1913 had a very short career. Seven orders placed with Stratford Works covered 51 locomotives. Beardmore & Co of Glasgow won an order for the 20 locomotives Nos 1541–1560, in 1920/21, while Beyer Peacock brought up the rear with its order for 10 from the LNER in 1928.

Construction of the new GER 4-6-0s continued at a steady rate between 1911 and the outbreak of World War I, when due to the demands of war effort work on available

workshop space the company went to an outside supplier. W. Beardmore & Co of Glasgow was awarded a contract to build 20 locomotives which were allocated order No W82 in Stratford's order book, and which were delivered in 1920/21. Meanwhile, a final batch of ten locomotives was ordered from Stratford and delivered between March and June 1920, even before Beardmore had completed its order. The final 4-6-0s to be delivered to the Great Eastern arrived, officially, from Beardmore in April 1921.

New Locomotives, Rebuilds and Modifications

After the 1923 Grouping one of the first changes to take place was the renumbering to 8500–8570.

In June 1926 No 8509 was fitted with a feedwater heater, with a feed pump from Worthington-Simpson and heater mounted on the left side of the footplate.

The following year, with the LNER pursuing further experimental work in this field, French-designed (ACFI) feed water heating apparatus was installed on no fewer than 50 locomotives. The LNER persisted with this work for some 10 years between 1927 and 1937, and although the company's attempts to reduce costs in this way were laudable, they were not a success. The maker's claims of reduced fuel costs for the ACFI installations were almost outweighed by the increased maintenance costs of the system. In

Outline diagram of Gresley rebuild of GE 4-6-0 as Class B12/3

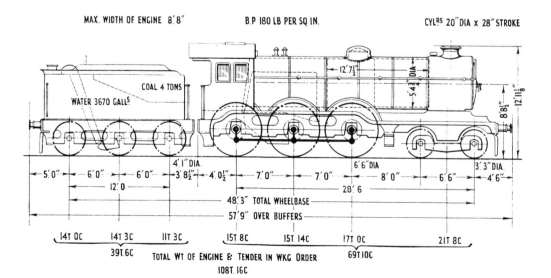

GER Class S69/1500 (LNER Class B12) 4-6-0 – Building and withdrawal

Original No	1924 No	1946 No	Built	Withdrawn	Scrapped
Built at Stratford Works					
1500	8500	1500	12/1911	6/1948	
1501	8501	1501	2/1912	5/1953	
1502	8502	1502	2/1912	4/1954	
1503	8503	1503	3/1912	5/1951	
1504	8504	1504	5/1912	6/1950	
1505	8505	1505	2/1913	3/1952	
1506*	—	—	2/1913	7/1913	
1507	8507	1507	3/1913	2/1953	
1508	8508	1508	3/1913	4/1953	
1509	8509	1509	4/1913	10/1948	
1510	8510	1510	4/1913	6/1949	
1511	8511	1511	5/1913	5/1952	
1512	8512	1512	6/1913	1/1957	Stratford
1513	8513	1513	6/1913	2/1953	
1514	8514	1514	6/1913	10/1959	Stratford
1515	8515	1515	11/1913	11/1951	
1516	8516	1516	11/1913	7/1958	Stratford
1517	8517	1517	11/1913	10/1948	
1518	8518	1518	12/1913	12/1947	
1519	8519	1519	12/1913	12/1957	Stratford
1520	8520	1520	4/1914	6/1957	Stratford
1521	8521	1521	4/1914	7/1952	
1522	8522	1522	4/1914	8/1947	
1523	8523	1523	5/1914	3/1955	
1524	8524	1524	5/1914	11/1953	
1525	8525	1525	6/1914	8/1951	
1526	8526	1526	6/1914	10/1951	
1527	8527	1527	8/1914	9/1947	
1528	8528	1528	8/1914	7/1953	
1529	8529	1529	9/1914	2/1950	
1530	8530	1530	11/1914	11/1959	Stratford
1531	8531	1531	11/1914	11/1947	
1532	8532	1532	12/1914	7/1953	
1533	8533	1533	12/1914	11/1959	Stratford
1534	8534	1534	2/1915	6/1945	
1535**	8535	1535	3/1915	12/1959	Stratford
1536	8536	1536	5/1915	12/1949	
1537	8537	1537	7/1915	4/1957	Stratford
1538	8538	1538	7/1915	1/1957	Stratford
1539	8539	1539	6/1917	11/1954	
1540	8540	1540	7/1917	10/1957	Stratford
Built by William Beardmore & Co Ltd					
1541	8541	1541	6/1920	1/1957	Stratford
1542	8542	1542	7/1920	7/1958	Stratford
1543	8543	1543	7/1920	6/1953	
1544	8544	1544	7/1920	9/1947	
1545	8545	1545	8/1920	1/1957	Stratford
1546	8546	1546	9/1920	5/1959	Stratford
1547	8547	1547	9/1920	10/1958	Stratford
1548	8548	1548	5/1921	12/1946	
1549	8549	1549	6/1921	1/1959	Stratford
1550	8550	1550	10/1920	1/1957	Stratford
1551	8551	1551	10/1920	1/1947	
1552	8552	1552	12/1920	7/1952	
1553	8553	1553	12/1920	8/1958	Stratford
1554	8554	1554	1/1921	9/1958	Stratford
1555	8555	1555	1/1921	10/1957	Stratford
1556	8556	1556	2/1921	12/1957	Stratford
1557	8557	1557	2/1921	1/1957	Stratford
1558	8558	1558	2/1921	4/1959	Stratford
1559	8559	1559	2/1921	9/1951	
1560	8560	1560	4/1921	5/1952	

Built at Stratford Works

1561	8561	1561	3/1920	9/1958	Stratford
1562	8562	1562	4/1920	8/1955	
1563	8563	1563	4/1920	4/1953	
1564	8564	1564	4/1920	11/1958	Stratford
1565	8565	1565	5/1920	1/1957	Stratford
1566	8566	1566	5/1920	1/1959	Stratford
1567	8567	1567	5/1920	11/1958	Stratford
1568	8568	1568	6/1920	8/1959	Stratford
1569	8569	1569	6/1920	1/1957	Stratford
1570	8570	1570	6/1920	3/1958	Stratford

Built by Beyer Peacock & Co Ltd

—	8571	1571	8/1928	12/1959	Stratford
—	8572	1572	8/1928	9/1961	Preserved
—	8573	1573	8/1928	1/1959	Stratford
—	8574	1574	8/1928	1/1957	Stratford
—	8575	1575	9/1928	4/1959	Stratford
—	8576	1576	8/1928	1/1959	Stratford
—	8577	1577	9/1928	9/1959	Stratford
—	8578	1578	9/1928	1/1957	Stratford
—	8579	1579	9/1928	1/1957	Stratford
—	8580	1580	10/1928	3/1959	Stratford

Notes

* No 1506 was withdrawn after being damaged beyond repair after an accident at Colchester in July 1913.

** No 1535 was ordered as replacement for No 1506.

In a partial re-numbering scheme in 1942, the B12s were allocated Nos 7415—84. Only the following were dealt with before the scheme was suspended and they reverted to their 85XX series:

85XX No	74XX No	85XX No	74XX No
8523	7437	8562	7476
8535	7449	8565	7479
8553	7467	8568	7482
8556	7470	8574	7488
8558	7472	8577	7491

In the BR numbering, 60,000 was added to the 1946 numbers.

GER Class S9/1500 (LNER Class B12) Building details

Stratford Order No	Number built	Running Nos	Where built	Works Nos
S69	5	1500–4	Stratford	–
A73	10	1505–14	Stratford	—
E75	5	1515–9	Stratford	—
R75	10	1520–9	Stratford	—
M77	6	1530–5	Stratford	—
B78	5	1536–40	Stratford	—
W82	20	1541–60	W. Beardmore & Co, Glasgow	135–41/3/4/42/5–54
H82	10	1561–70	Stratford	—
(1513*)	10	8571–80	Beyer Peacock, Manchester.	6487–6496

* Beyer Peacock order number

LNER Class B12 Diagram numbers

Diagram	Class	Details
Section E, 1924	B12	Belpaire firebox fitted.
Section E, 1926	B12	Fitted with Lentz poppet valve gear.
Section LNE, 1922	B12/2	Beyer Peacock locomotives; classification altered in 12/1929.
Section LNE, 1931	B12/2	Locomotive reconverted to piston valves, from Lentz valve motion.
Section LNE, 1932	B12/3	Locomotives rebuilt with 5 ft 6 in diameter boiler. Replacement diagram. Gives details of the reduction in boiler tubes.
1946	B12/4	Diagram 25A boiler, round-topped firebox, and classification altered to B12/4 in 12/1948.

that aspect the variable and predominantly hard water with which the B12s operated caused substantial furring of the apparatus. All locomotives fitted with the equipment had it removed by the end of the 1940s. The ACFI equipment consisted of large drums carried on top of the boiler, and a tandem steam pump at the smokebox end on the running boards, on the left side. The two pump cylinders were operated by a central steam cylinder, with one of the pumps taking water from the tender to one of the boiler mounted reservoirs, where it was mixed with exhaust steam in a fine spray, and allowed to drain into a sump, and on into the second reservoir. The second pump lifted the water from this second reservoir into the boiler.

In 1926, the LNER fitted No 8516 with Lentz oscillating cam poppet valve gear. It was only the second locomotive in the country to have this gear, and it was an experimental fitting to determine if the better steam distribution and fuel economies claimed by the makers were true. In the previous year J20 0-6-0 No 8280 had been the first with this gear, but in the B12 installation modifications and improvements were made. Inlet and exhaust valves were $6^5/_8$ in and 7 in respectively, with the cams actuated from the existing Stephenson valve gear, and screw reverse was provided, compared with air-operated reverse on normal B12s. The initial results obtained, with good acceleration and point-to-point timings and performance, was generally found to be more economical, particularly on short or intermediate stopping duties.

This promise of improved performance with the Lentz valve gear on No 8516 resulted in the same equipment being provided on No 8525 in 1928, and the Beyer Peacock order for ten new B12s was amended to include the Lentz poppet valve gear. However, faults developed, some of which were attributable to the design of the mechanism, with twisted camshafts and cracks in the camboxes. In the B12s, the camboxes were cast integrally with the cylinders; consequently the nature of the faults demanded early and costly renewal of cylinders on the locomotives affected. With such major problems occurring it was decided to rebuild the Lentz fitted locomotives from 1931, with some of the rebuilds immediately receiving new, long-travel valves, so recently

initiated by Stratford Works.

Other modifications arrived with the batch of Beyer Peacock B12s in 1928, which were originally known as B12/LNE, a classification which was changed to B12/2 in 1929. They were also outshopped with 1924 series running numbers, from 8571 to 8580. The most important mechanical alterations included deeper frames between the driving axle and cylinders, the abandonment of the sideways moving rear coupled axleboxes, and a reduction in the number of small tubes in the boiler to 165. Further changes on the Beyer Peacock locomotives included an extended smokebox, cut away at the base, and cast-iron chimney. The alterations to the boiler significantly reduced the heating surface, which came down to 1,481.3 sq ft; although initially known as diagram 99, they were later included with the diagram 25 boilers on other B12s. The firebox was still of the Belpaire type, but the firehole door was changed to the former GER oval pattern, and the four column Ramsbottom safety valves were replaced by the Ross pop variety.

Other differences noted on the B12/2s included the removal of the ornate valance, and changes in the sanding arrangements, together with a hand-operated water scoop mechanism. The maximum axle loads on the new locomotives went up to 15 tons 17 cwt.

Rebuilds
There were two major rebuilds of the class during LNER days, the B12/3 and B12/4 versions, with the first of these taking place in the early 1930s, and involving larger boilers and a redesign of the front end. The first to receive the alterations was No 8579 in May 1932, when a 5 ft 6 in diameter boiler was fitted, similar in arrangement to that used on the B17 Sandringham class. The B12s had a shorter barrel than the B17s, with other differences in the firebox design, which now featured the Gresley round-top arrangement. A larger superheater was installed, with 24 elements. Modifications were made to the Stephenson valve gear. The new long-travel valves with a maximum travel of $6^1/_{16}$ in were used, but with less steam lap and $9^1/_2$ in diameter valves.

Rebuilding and modifications saw the disappearance of the Lentz valve gear, ACFI feedwater heating apparatus. By the early 1940s only 16 of the original B12s remained in stock. The majority by that time were much changed, substantially more like Gresley locomotives than those of Holden.

In 1943 a second rebuilding, or partial rebuilding, affected nine B12s (including the first locomotive, No 1500) all of which by 1948 had become Class B12/4. The changes involved the Scottish locomotives, where restrictions on axle loading had previously prevented the use of further rebuilds. A redesign of LNER boilers to new standards was instigated in 1941, just as the boilers on the Scottish B12s were coming due for replacement. In this redesign the Belpaire firebox disappeared and the barrel was rolled from a single plate, replacing the previous two ring design. The arrangement of tubes was altered to give a pattern of diagonal instead of horizontal rows. These were known as diagram 25 boilers, and were carried by Nos 1500/11/24/2632, 8504/5/7/8.

Modifications

The obvious changes in the appearance of the B12s resulting from the use of four different boiler designs were supplemented by a number of other modifications. Among these minor changes made in LNER days were trials of various forms of tube cleaner, soot blowers, patent speed-recording devices and cut-off control equipment, and the fitting of a Kylala blastpipe on No 8519 in July 1927.

Locomotives which were rebuilt to class B12/3 saw a number fitted with new frames in the late 1940s with the disappearance of the side traversing rear axlebox, while it was also removed on some locomotives which did not have new main frames. Cast-iron chimneys began to replace the original built-up variety. Soon after the Grouping the decorative valancing also began to disappear in a number of cases. Buffers and drawgear changed after 1923, with these becoming LNER group standard in place of the lighter GE pattern, although only some B12s received the new hardware.

The feedwater heating experiments which began in 1926 when No 8509 was fitted with the Worthington-Simpson apparatus and which became more extensive with the application of ACFI equipment, was removed from the B12/3 rebuilds. These were fitted with the 5 ft 6 in diameter diagram 99A boiler. The ACFI installations had resulted in minor modifications to pipework

runs, and some locomotives retained the altered layout when the apparatus was removed.

Mechanical lubrication of the front end was standard for B12s, but with a variety of different makes of equipment. Nos 1561–1568 had Detroit sight feed lubricators, with axleboxes lubricated by syphon feed. Ten locomotives had the Wakefield Fountain type, while another ten had Empire twin plunger lubricators. The Wakefield type survived the rebuilding to B12/3, but the Empire type disappeared quite quickly.

Tenders

The standard type of tender fitted to the B12s were 3,700 gallons capacity, and held four tons of coal. No 8578 received the tender originally built for the 'Decapod', following a collision in January 1931. The B12 tenders were 18 ft 4$\frac{1}{2}$ in long over the frames, with 4 ft 1 in wheels in a wheelbase of 12 ft 0 in equally divided. The wheels protruded into the water tank, with small splashers covering the intrusion. On top of the tank balancing pipes were fitted. Coal guards were added by the LNER, except for the Beyer Peacock locomotives which had them from new. Originally the water pick-up gear was air-operated, although this was removed on the B12/3 rebuilds; the Beyer Peacock locomotives were provided with mechanically operated gear.

Tablet exchange apparatus was installed on locomotives working the Midland & Great Northern and the Great North of Scotland sections, while some of the Scottish locomotives acquired B17 tenders after World War II. The earlier Sandringhams were intended as replacements for the B12s in East Anglia, and as a result the tenders of the original series held only 3,700 gallons of water, and five tons of coal.

Operations

In Great Eastern days Stratford and Ipswich saw the greatest concentration of B12s, which were set to work on the main lines to Colchester and Norwich, and on boat trains to Parkeston Quay. By 1921/22 the principal expresses from London were seeing the new 4-6-0s, while the introduction of the new Hook Continental service saw continued success for the B12s in LNER days. Also under the LNER's guidance, trials were conducted

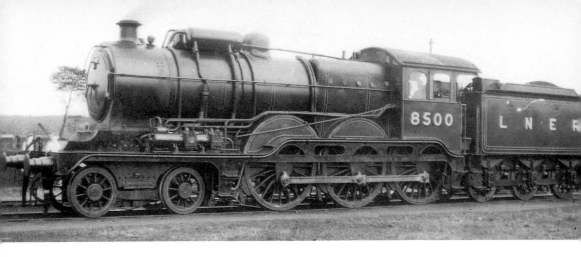

Gresley's boiler feedwater heating experiments took a number of forms on the LNER, in the case of the ACFI apparatus installed on the B12s. Seen on No 8500, it gave rise to the nickname 'Hikers' in Scotland, or 'Camels' in East Anglia. *L&GRP*

with the class in service in the Leeds–Doncaster area, though the locomotives involved soon returned to the GE section. In 1926 tests were conducted with No 8526 over former GNofS routes in Scotland, with the result that the 4-6-0s proved ideal replacements for existing 4-4-0 types. Trials were conducted on other LNER routes, including ex-GCR lines, although it was not until the late 1920s that the Great Eastern section motive power shortage saw some improvement.

The arrival of the first Class B17 Sandringham 4-6-0s in 1931 enabled the transfer of B12s to Scotland, where they were initially employed on the Aberdeen to Elgin route. The ACFI-fitted B12s were known as 'Hikers' in Scotland, with similarly equipped locomotives on the GE Section acquiring the nickname 'Camels'. Initially on the GNofS main lines they worked goods and fish trains, but soon migrated to a variety of passenger duties, and in the summer months Kittybrewster would loan B12s to Eastfield for excursion workings over the Oban line. Some trials were carried out with Edinburgh based locomotives on Glasgow expresses, but their performance was reportedly inadequate, due to steaming difficulties. The class was retained on the GNofS section, and those that had been transferred to other regular duties in Scotland were sent to the GNofS lines, under the guise of being deficient in brake power.

South of the Border, the B12s were supplemented on passenger work by increasing numbers of B17s, while the rebuilt B12s earned reputations as more powerful, fast-running machines. In the 1930s, the heavy Hook Continental, reaching a weight of around 465 gross tons, was still rostered B12/3s. At that time most boat train workings were in the hands of the Sandringham class locomotives, which were also proving popular on the longer Norwich workings to and from London. World War II saw the B12s putting in sterling service on very heavy loadings, often reaching up to almost 500 tons. They were an equally common sight at the head of ambulance trains which comprised air-braked stock brought over from the USA, and which necessitated a minor alteration to the proportional valve fitted to the B12/3s used.

Following the end of the war, when the motive power position on the GE Section changed with more B17s and an influx of the new Thompson B1 4-6-0s the B12s tended to be concentrated at Stratford. From this depot, the semi-fast Southend trains were almost their exclusive role. Of the original unrebuilt B12s Nos 8521/3, which had returned to East Anglia for a short time in 1944, joined the GE Section No 8534 at Ipswich. The latter was withdrawn in June 1945, and the last remaining unrebuilt locomotives on the GE Section were transferred back to Scotland in 1945.

Just before nationalisation, the concentration of B12s revealed that of the 50 B12/3s still at work in East Anglia 37 were based at Stratford, four at Colchester, seven at Ipswich, and one each at Norwich and Yarmouth (South Town). Changes in their area of operations did take place in BR days, along with a number of minor modifications.

Although the class did not survive the 1950s, No 61572 was rescued for preservation, and was being restored by the North Norfolk Railway in 1988.

CHAPTER FIVE

THE GRESLEY YEARS

With the grouping of the railways in 1923 the newly-formed LNER acquired a number of 4-6-0 designs, mainly from the Great Central, North Eastern, and Great Eastern railway companies. At the same time, the man appointed as chief mechanical engineer of the new company was actually second choice, and came from the Great Northern Railway, which had no examples or real experience of the 4-6-0 type. Robinson declined the cme's post on account of his age; Gresley went on to become one of this country's most eminent locomotive engineers. Unlike Robinson, who had produced a number of 4-6-0s for the Great Central, Gresley earned his reputation designing Pacific types, although introducing many successful mechanical details applied to other types.

In the mixed-traffic category there was no policy of scrap and build, and standardisation on a few locomotive designs. The LNER was not a wealthy company, and the Gresley era was characterised in a number of ways by the development and improvement of existing locomotive types. Most of the inherited 4-6-0s were subject to modification and alterations, as outlined in previous chapters. In new locomotives, Gresley introduced only one 4-6-0 design, the B17 Sandringham class, intended for the Great Eastern lines, where the severe weight restrictions precluded the use of Pacific types. The design was later subject to alteration, initially only cosmetic, with two examples sporting the streamlined casing of the form used on the Pacifics, while more radical changes were introduced by Edward Thompson. The later changes introduced by Thompson included a move away from Gresley's favoured three-cylinder drive and a new design of boiler. Ultimately the alterations to the Sandringhams did not show any marked improvement in performance or economy over the original version, and the policy was not continued.

THE SANDRINGHAMS – CLASS B17
Having said that Gresley's new 4-6-0 was

produced in response to the requirements of the GE section, it was in fact a design produced rather more by the North British Locomotive Co than the LNER. Despite the pressing need of the Eastern Section routes for greater locomotive power than the existing B12s were able to provide, the restricted axle loadings imposed severe limitations on the design. In addition, given the size of available turntables, the overall length of the locomotives was an important consideration. The maximum combined coupled axle load laid down for the GE section was 44 tons, and this was governed by the strength of bridges along these routes.

What eventually arrived in May 1928 as the new B17 Sandringham class could trace its ancestry to a 2-6-4 tank engine design, abandoned after the accident which befell the Southern Railway River class 2-6-4 passenger tank in 1927. Gresley had developed a design with 6 ft 2 in coupled wheels, 20 in by 26 in cylinders, and a J39 boiler, which after the derailment on the Southern was dropped in favour of another ten B12s from Beyer Peacock, with long travel valve gear.

Findings by the Bridge Stress Committee showed that axle loading alone was not the only parameter, but demonstrated hammer-blow to be of even greater importance. Using Gresley's three-cylinder drive, with weight-saving techniques in construction like the use of alloy-steel motion parts, it was found possible to raise the maximum coupled axle load on the proposed new 4-6-0 for the GE section to 54 tons. Although Gresley began work on the new design, other matters attracted his attention, and the detail work was handed over to the North British Locomotive Company.

The scheme put forward by the LNER at this stage was of unsuitable weight distribution, with the drive for the inside cylinder provided in the usual Gresley manner, and taken from the middle coupled axle. The way in which the weight distribution problem was overcome was to divide the drive, with the two outside cylinders driving onto the centre pair of coupled wheels, and the inside cylinder driving the leading pair. The loading on the centre axle was kept down to 18 tons, with 18 tons on the leading pair, and 17 tons 15 cwt on the rear. The bogie supported no less than 22 tons 18 cwt.

On their first appearance, as with many new designs, there were teething troubles, although these were soon overcome, and the locomotives settled down to become one of the most successful 4-6-0 types. Embodying all that was typical of the Gresley era, mechanically at least, it was perhaps only in their complexity that they were later at a disadvantage. By the 1940s locomotive design had moved on to more simple standardised designs and Thompson's Class B1 4-6-0 successfully gave the LNER a well-liked general purpose locomotive. The B17s were also well liked by footplatemen and enthusiasts, although neither a Sandringham, nor one of the later 'Football Club' series has been rescued for preservation.

Boiler Design

The new 4-6-0s for the GE section were fitted with a large parallel boiler, built in three rings, and paired with a round-topped fire-

One of the earlier B17s, No 2834 *Hinchingbrooke*, piloted by ex-GCR B3 4-6-0 No 6169 *Lord Faringdon* at Sheffield Victoria station. *Stephenson Locomotive Society*

box. In original form the length between tubeplates was 14 ft 0 in, with the regulator housed in the steam dome on the central ring; the outside diameter under the lagging, was 5 ft 6 in. The original boiler pressure was to have been 180 lb/sq in, a standard for the period, but it was increased to 200 lb/sq in during the design stage. The problem that presented itself was one of a reduction in the ability of the boiler plates to present a sufficient safety margin at the higher pressure after a period of wear and corrosion. In later life it was necessary to revert to the original pressure when the older boilers were around 15 years of age, during World War II.

The boiler plates were $5/8$ in thick, with $1/2$ in thick wrapper plates and an outside diameter of 5 ft $9\frac{1}{2}$ in. The smokebox of the same diameter was the drumhead pattern, 5 ft $5\frac{1}{2}$ in long and supported on a cast-steel saddle. The small boiler tubes were 2 in diameter, and were 12 swg thick. The large $5\frac{1}{4}$ in diameter superheater flues contained twenty-four $1^7/_{32}$ in elements. The Robinson style superheater was a standard fitting, with the header in the smokebox incorporating a single snifting valve, seen behind the locomotive chimney.

The draughting was initially not satisfactory, and resulted in a series of experiments with various designs of blastpipe, varying from the original $5\frac{1}{4}$ in diameter cap, positioned just $3\frac{1}{2}$ in below the boiler centre

LNER Class B17 Sandringham 4-6-0 – Leading details

Designer	H. N. Gresley/North British Locomotive Co		
Built	12/1928–7/1937		
Withdrawn	9/1952–8/1960		
Running numbers			
– original	2800–2872		
– 1946	1600–1672		
– BR	61600–61672		
Wheel diameter			
– coupled	6 ft 8 in		
– bogie	3 ft 2 in		
Wheelbase	48 ft 4 in		

Axle load	Tons Cwt	Tons Cwt	Tons Cwt
– coupled	18 00	18 00	17 15
– bogie	22 18		

Boiler – type	100	100A*
– diameter	5 ft 6 in	5 ft 6 in*
– length	14 ft 0 in	13 ft 11⅞ in*
– tubes: small	143 × 2 in o/d	143 × 2 in o/d*
– tubes: large	24 × 5¼ in o/d	24 × 5¼ in o/d*
Heating surface		
– tubes	1048.00 sq ft	1033.00 sq ft**
– firebox	168.00 sq ft	168.00 sq ft**
– flues	460.00 sq ft	460.00 sq ft**
– Total	1676.00 sq ft	1661.00 sq ft**
Superheater elements	24×1¾ in o/d	24×1.244 in o/d*
Superheater	344 sq ft	344 sq ft
Working pressure	200 lb/sq in (original)	200 lb/sq in
Grate area	27.5 sq ft	27.9 sq ft
Cylinders		
– number	3	
– dimensions	17½ in × 26 in	
Tractive effort	25,380 lb (original)	
Adhesive weight	Between 53 tons 15 cwt and 56 tons 10 cwt	
Fuel capacities		
– coal	4 tons with GE pattern tender/7½ tons with LNER standard	
– water	3,700 gallons with GE tender/4,200 gallons with LNER pattern	

Weights in working order	GE tender	LNER tender
	Tons Cwt	Tons Cwt
– locomotive	76 13	76 13
– tender	39 08	52 00
– Total	116 01	128 13

* Diagram 100A boiler was fitted to Class B17/6
** These changes were made to the number of tubes/heating surface of diagram 100 boilers, from 11/1945 onwards.

line. Experiments lasted from the late 1920s until the mid 1930s, and involved changes to the diameter and internal shape of the blast-pipe top, eventually resulting in the fitting of a 5⅛ in diameter 'flat bridge' type.

Tests with the new design in 1933 proved successful in improving the boiler's steaming ability, and reducing the lifting effect on the fire. In addition to being adopted as standard B17 fittings, the flat bridge design of blast-pipe was standard on Gresley Pacifics. Only a year later another experimental blastpipe produced even better results, with a sharper blast – the constant acceleration blastpipe cap. This was then adopted as standard for the Sandringhams, and positioned 8¼ in below the boiler centre line, although the flat bridge type was fitted to many (and on some Darlington built locomotives), they were only 5 in diameter and 7¼ in below the boiler centre line.

At the rear, the standard Gresley round-topped firebox was fitted, with steel outer plates and a copper inner firebox, with a maximum outside length of 10 ft 0½ in and a width of 4 ft ½ in. Overall, construction was generally lighter than contemporary practice, all in the interest of saving weight, with wrapper plates only ½ in thick. The copper inner firebox was fabricated from 9/16 in thick plates for the sides, and 1 in thick for the tubeplates, providing a total heating surface of 168 sq ft. The fire grate, with an area of 27½ sq ft, had a sloping front half and a flat rear section over the rear coupled axle. The front portion of the grate was the Gresley drop type, and on Nos 2800–2809 was operated by a screw mechanism and complex series of rods, cranks and levers between the frames. From March 1933 simplified operating gear was installed on Nos 2810 onwards, with the screw mechanism acting on a shaft connected directly with the drop section of the grate, and clearly visible on the right-hand side of the locomotive. The earlier locomotives were fitted later with this arrangement. The grate area of the B17s was only 1 sq ft more than that of the B12s which they were replacing, although the overall heating surface was significantly greater.

Frames, Wheels and Motion

Construction of the main frames for the B17s, although involving conventional materials – 1⅛ in thick mild steel plate – adopted methods of lightening the frames' structure, and arrangement of stays which soon proved troublesome in service. Frame fractures were an early fault in the first batch built by the North British Locomotive Company, and resulted in the fitting of new front end frames to almost all of the first Sandringhams, following attempts in 1931 at the fitting of stiffening plates. Hornblocks were only fitted originally to the driving axleboxes, with simpler guides installed in the centre and trailing horn gaps. Later, it was found necessary to provide hornblocks to the centre axle, as the fractures in the vicinity of the horn gaps continued through the 1930s. Hornblocks continued to cause the B17s problems, with a high level of replacement, and during World War II a shortage of steel castings led the LNER to order more from the LMS, due to their pressing need.

Later batches of B17s, along with those built at Darlington, required less drastic modifications to their original frame layout, although the horn gap fractures remained a problem. Further modifications to the springing, partly intended to offset complaints of bad riding, included reducing the thickness of the plates to $9/16$ in. The nine $5/8$ in plates in the original design were replaced by one $5/8$ in plate and twelve $9/16$ in plates. The severity of the weight limits imposed on the East Anglian routes was a contributory factor in the frame fractures that occurred in service. It is perhaps ironic that the BR Standard Pacifics which were also first put to work on the GE section also suffered frame

fractures, among other troubles.

Locomotive main axleboxes were plain bearing type, although Gresley had been introducing roller bearings in a limited way on the pins of the valve gear, following almost 10 years of service experience on a former GNR 2-6-0. The journals of the new B17s were $9^{1}/_{2}$ in by 11 in, with the 6 ft 8 in diameter coupled wheels having 20 spokes. In later life, the axleboxes of the B17s, particularly the trailing sets, developed excessive sideplay, with a knock-on effect on the quality of ride. Similar but less pronounced wear was demonstrated in the leading and intermediate journals, although the complaints were more prevalent for the trailing pair. The strain of the additional workloads imposed by World War II had some effect on the general wear and tear in these locomotives, as it had on many others.

The North British Locomotive Co, from whom came the design in collaboration with the LNER design team, had produced a locomotive just over the original specification for the maximum axle load of 17 tons. In this first series, 18 tons rested on the leading and intermediate axles, and $17^{3}/_{4}$ tons on the rear. The Football Club series of 1936/37 was heavier on all coupled axles – 18 tons to the front and rear coupled axles, and 18 tons 7 cwt on the intermediate pair. The coupled wheels all had laminated springs. The leading bogie was of simple plate frame construction, with helical springs and 3 ft 2 in diameter wheels running in $6^{1}/_{2}$ in by 11 in journals.

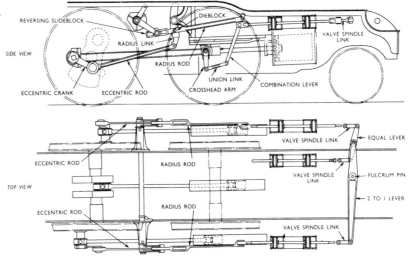

Gresley's use of Walschaert's valve gear with derived linkage for the middle cylinder valve

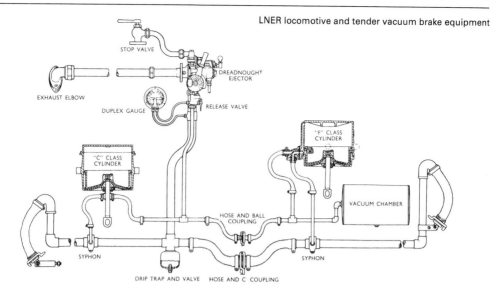

LNER locomotive and tender vacuum brake equipment

According to contemporary technical press reports, the three cylinders were operated by two sets of valve motion, Walschaerts valve gear for the outside cylinders, and the Gresley derived motion for the inside cylinder. The conjugate valve gear, where movement of the inside cylinder valves was obtained from the outside gear, was a trademark of Gresley design. In the Sandringhams, the 2-to-1 lever was placed behind the cylinders. This layout had caused some problems in the design stage, and demanded that the inside cylinder be placed well forward in the frames, driving onto the leading axle. The 8 in diameter piston valves were carried above the cylinders, and had a maximum travel of $5^5/8$ in and a cut-off in full gear of 65%, with a calculated cylinder horsepower of 1,653. The boiler was quoted as 1,230 hp at the working pressure of 200 lb/sq in. During trials held in 1936 with the LNER dynamometer car No 2861 *Sheffield Wednesday* recorded a highest output of 1,400 hp at 45% cut-off. Other sources list the maximum valve travel as $5^{21}/32$ in. No 2803 was initially fitted with a patent cut-off control gear, later removed. Lubrication of the cylinders and valves was achieved by means of two different designs of Wakefield mechanical lubricator.

There were three types of brake gear on the Sandringhams; steam, vacuum, and Westinghouse air brake. The last was installed originally on the first 16 Sandringhams, for locomotive and train, with the $8^1/2$ in compressors mounted on the right-hand side of the smokebox, and operating cylinders between the frames. Curiously perhaps, the Westinghouse couplings were carried below the front buffer beam as on former NER classes and the LNER D49 class 4-4-0s, but were later re-positioned above the buffer beam, as requested by the GE section authorities. Nos 2800–2815 were also fitted with a vacuum ejector for working vacuum fitted stock, with a large reservoir carried between the driving and intermediate axles, inside the frames.

Locomotives built at Darlington, numbered from 2816 to 2842, were fitted with steam brakes on the locomotive and vacuum for the train; a policy change in 1933 saw vacuum brakes standardised for the rest of the class. Some of the vacuum-braked locomotives were later fitted with Westinghouse air brakes when B12s were transferred to Scotland, while the Westinghouse fitted locomotives had the equipment removed when a number were rebuilt as Class B2 in Thompson's day. Throughout their later operating careers with the LNER the brake systems fitted to Sandringhams were causing some confusion as to which locomotives were fitted with which equipment.

Superstructures

The Sandringham was an elegant design in the true Gresley mould, with the running boards curving gracefully over the coupled wheels, and the narrow valance an unobtrusive feature of their appearance. At the front of the locomotive, the main frames protruded over the front step plate in a gentle curve. Footsteps were conspicuous by their absence in early examples of the class; the narrow footsteps alongside the cab were added later. Since only one was at first required, the width of the running board was increased by 1½ in. In the North British locomotives nameplates were attached to the leading splasher on the first four, but these were immediately repositioned by the LNER on the centre splasher which was made some 4 in wider and bolted to the running board rather than rivetted down, in order to facilitate positioning of the reach rod.

The overall width of the B17s at 8 ft 7 in was the same as that of the B12s, but in the cab standard Doncaster practice with a well in the centre of the floor proved troublesome. In the B12s the firehole was only 1 ft above floor level, but in Gresley's new design the wooden floor with its central well was more than double this height from the firehole, making firing difficult and uncomfortable. The solution on the GE section locomotives Nos 2810–2847 was simply to fill-in this well with a wooden decking.

The introduction of the standard LNER tenders demanded a further alteration, since the shovelling plate of these was around 1 ft lower than in the old GE variety. The wooden cab floor was made to slope upwards to the firehole on Nos 2848–2872 – more difficulty with firing!

Locomotives from No 2837 onwards were built with hinged sight screens between the side windows; the earlier examples were retrospectively fitted.

Some locomotives were also provided with speed indicators/recorders, but it was not common, and restricted to fewer than half a dozen B17s before World War II.

Minor alterations were quite numerous for both the earlier locomotives on the GE section and the later examples put to work on former Great Central metals. Among these changes can be included the variations introduced as the LNER standardised its design and construction practices. Changes in the design of lamp irons was one example of this, while most B17s had the boiler handrail stanchions placed further away from the cleading, to prevent cleaners' boots from damaging the paintwork. Handrails were fitted to the front corners of the cab on a number of locomotives. Most sported twin guard irons on the bogie and locomotive main frames until 1951, when, following changes in maintenance practice, the irons attached to the locomotive frame were removed.

The axlebox oil feed pipes were a prominent feature on the boiler side, leading from the lubricator in the cab to each coupled wheel position. Sandpipes were another item which varied, with one or two members of the class having steam sanding on the centre coupled wheels, and others with gravity feeds to the rear of the trailing pair.

At least one member of the class, No 61653 *Huddersfield Town*, ran with an A3 type chimney for some time instead of the 1 ft 0¼ in pattern normally fitted.

The locomotives were provided with extended cylinder drain pipes from No 2811 onwards, with the rest of the class altered between 1936 and 1938.

The most striking external change was the streamlining of Nos 2859 and 2870 in September 1937 for working the East Anglian service. These two locomotives were classified B17/5. The streamlining was essentially a cut-down version of that fitted to the Class A4 Pacifics, with only minor alterations to the locomotives themselves. A cut-out in the casing, in front of the trailing coupled wheel, exposed the Wakefield lubricator, while the cab spectacle plate was angled to the casing. As with the Pacifics a deep valance covered much of the wheels and motion, although during World War II this was removed to give easier access for maintenance and repair.

At the leading end of the streamlined locomotives the steam pipes protruded through the casing slightly, while the buffers were extended by some 7 in compared with a standard B17. Also like the streamlined Pacifics a chime whistle was carried, mounted in front of the chimney. The tenders were the larger 4,200-gallon LNER standard type, with additional streamlined fairings, while a flexible rubber shroud covered the connection between locomotive and tender. These were at the time the only LNER standard tenders on the former GE lines, while the streamlining raised the weight in working order from 77 tons 5 cwt to 80 tons 10 cwt.

Class B17 No E1664 *Liverpool* seen in February 1948 in the early BR livery with the full title on the tender, BR and E prefix to the running number. *H. C. Casserley/Lens of Sutton*

Construction

In total 73 were built, ten from the North British Locomotive Co, ten from Robert Stephenson & Hawthorn, and the remainder from Darlington Works. The first 27 of the Darlington series which appeared in 1930 and 1931 had boilers supplied by W. G. Armstrong Whitworth & Co. Originally too, because of the varied suspension arrangements, the class was divided into four subclasses, although from the late 1930s the first three of these were amalgamated, since they all had the small GER pattern tender. The B17/4 locomotives were numbered 2848 to 2872, and were paired with LNER standard tenders from new.

The Sandringhams were constructed in seven batches and were needed urgently, following disastrous operating problems on the Cambridge and Southern routes during the winter of 1926/27. The civil engineer's restrictions on axle loadings posed a number of problems in the design phase at Doncaster, and later at the North British Locomotive Company's works, to whom the first contract was awarded. Attempts had been made at Doncaster to derive a new 4-6-0 with three cylinders, driving the leading axle, using the same cylinder and motion layout adopted for the D49 class 4-4-0s, and the general layout of the former NER B16 4-6-0.

The initial tender for the B17s was for 20 locomotives, but only ten were accepted by the LNER, and the contract was awarded on 17 February 1928, with each locomotive costing £7,280, to be delivered between August and November that year. North British had submitted two alternatives with its tender, and although the LNER accepted the 18 tons axle load of the lighter version, further alterations were required. Order No L850 covered these ten B17s, in which a number of components of the design, notably the cab, cylinders and motion, were obtained from drawings of the Pacific types, which the company had built four years earlier. The boiler, too, owed something of its ancestry to the LNER K3 2-6-0 and O2 2-8-0.

Problems were again encountered in the design process soon after the contract was awarded, particularly in respect of the drive, and NBL finally settled the matter by dividing the drive in a similar way to that used on the Royal Scot 4-6-0s for the LMSR. The changes which the LNER required ranged from an increase in the cylinder diameter to $17\frac{1}{2}$ in, lengthening the frames at the cab end, using lighter springs, and positioning the 2-to-1 lever behind the cylinders. The difficulties with the design led North British to anticipate thirteen weeks' delay in delivery, while the LNER Locomotive Committee had reported in October 1928 that the penalty clause in the contract with NBL had been removed. Presumably this was done in the light of the changes in the design demanded by the company.

LNER Class B17 4-6-0 – Building and withdrawal

Original No.	Name	Built	1946 No	Withdrawn *Rebuilt B2	Scrapped at
Built by the North British Locomotive Co Ltd. Works No. 23803–23812					
2800	Sandringham	12/28	1600	7/58	Doncaster
2801	Holkham	12/28	1601	1/58	Doncaster
2802	Walsingham	11/28	1602	1/58	Doncaster
2803	Framlingham	12/28	1603	*10/46	
2804	Elveden	12/28	1604	8/53	Doncaster
2805	Burnham Thorpe	12/28	1605	5/58	Doncaster
2806	Audley End	12/28	1606	9/58	Doncaster
2807	Blickling	12/28	1607	*5/47	
2808	Gunton	12/28	1608	3/60	Stratford
2809	Quidenham	12/28	1609	6/58	Doncaster
Built at Darlington Works					
2810	Honingham Hall	8/30	1610	1/60	Doncaster
2811	Raynham Hall	8/30	1611	10/59	Doncaster
2812	Houghton Hall	10/30	1612	9/59	Doncaster
2813	Woodbastwick Hall	10/30	1613	12/59	Doncaster
2814	Castle Hedingham	10/30	—	*11/46	
2815	Culford Hall	10/30	—	*4/46	
2816	Fallodon	10/30	—	*11/45	
2817	Ford Castle	11/30	1617	*12/46	
2818	Wynyard Park	11/30	1618	1/60	Doncaster
2819	Welbeck Abbey	11/30	1619	9/58	Doncaster
2820	Clumber	11/30	1620	1/60	Stratford
2821	Hatfield House	11/30	1621	11/58	Doncaster
2822	Alnwick Castle	1/31	1622	9/58	Doncaster
2823	Lambton Castle	2/31	1623	7/59	Doncaster
2824	Lumley Castle	2/31	1624	3/53	Doncaster
2825	Roby Castle	2/31	1625	12/59	Doncaster
2826	Brancepeth Castle	3/31	1626	1/60	Doncaster
2827	Aske Hall	3/31	1627	7/59	Doncaster
2828	Harewood House	3/31	1628	9/52	Doncaster
2829	Naworth Castle	4/31	1629	9/59	Doncaster
2830	Thoresby Park	4/31	1630	8/58	Doncaster
2831	Serlby Hall	5/31	1631	4/59	Doncaster
2832	Belvoir Castle	5/31	1632	*7/46	
2833	Kimbolton Castle	5/31	1633	9/59	Doncaster
2834	Hinchingbrooke	6/31	1634	8/58	Doncaster
2835	Milton	7/31	1635	1/59	Doncaster
2836	Harlaxton Manor	7/31	1636	10/59	Doncaster
2837	Thorpe Hall	3/33	1637	9/59	Doncaster
2838	Melton Hall	3/33	1638	3/58	Doncaster
2839	Rendelsham Hall	3/33	—	*1/46	
2840	Somerleyton Hall	5/33	1640	11/58	Doncaster
2841	Gayton Hall	5/33	1641	1/60	Stratford
2842	Kilverstone Hall	5/33	1642	9/58	Doncaster
2843	Champion Lodge	5/35	1643	7/58	Doncaster
2844	Earlham Hall	5/35	(61644)	*3/49	
2845	The Suffolk Regiment	6/35	1645	2/59	Doncaster
2846	Gilwell Park	8/35	1646	1/59	Doncaster
2847	Helmingham Hall	9/35	1647	11/59	Doncaster
2848	Arsenal	3/36	1648	12/58	Doncaster
2849	Sheffield United	3/36	1649	2/59	Doncaster
2850	Grimsby Town	3/36	1650	9/58	Doncaster
2851	Derby County	3/36	1651	8/59	Doncaster
2852	Darlington	4/36	1652	9/59	Doncaster
2853	Huddersfield Town	4/36	1653	1/60	Doncaster
2854	Sunderland	4/36	1654	11/59	Doncaster
2855	Middlesbrough	4/36	1655	4/59	Doncaster
2856	Leeds United	5/36	1656	1/60	Doncaster
2857	Doncaster Rovers	5/36	1657	6/60	Stratford
2858	Newcastle United	5/36	1658	12/59	Doncaster
2859	Norwich City	6/36	1659	3/60	Stratford
2860	Hull City	5/36	1660	6/60	Stratford
2861	Sheffield Wednesday	6/36	1661	7/59	Doncaster

LNER Class B17 4-6-0 – Building and withdrawal

Original No.	Name	Built	1946 No	Withdrawn *Rebuilt B2	Scrapped at
Built by Robert Stephenson & Co Ltd. Works Nos 4124—4134					
2862	Manchester United	1/37	1662	12/59	Doncaster
2863	Everton	2/37	1663	2/60	Doncaster
2864	Liverpool	1/37	1664	6/60	Stratford
2865	Leicester City	1/37	1665	4/59	Doncaster
2866	Nottingham Forest	2/37	1666	3/60	Stratford
2867	Bradford	4/37	1667	6/58	Doncaster
2868	Bradford City	4/37	1668	8/60	Stratford
2869	Barnsley	5/37	1669	9/58	Doncaster
2870	Manchester City	5/37	1670	4/60	Stratford
2871	Manchester City	6/37	—	*8/45	
2872	West Ham United	7/37	1672	3/60	Stratford

NOTES:
1 No 2844 was renumbered direct into BR series when rebuilt.
2 BR numbers — 1946 numbers, increased by 60,000.
3 Works numbers are in same sequence as running numbers.
4 Locomotives re-named:
    2805 Lincolnshire Regiment (4/38)   2870 (1) Tottenham Hotspur (5/37)
    2830 Tottenham Hotspur (1/38)   (2) City of London (9/37 — when streamlined)
    2839 Norwich City (1/38)   1671 Royal Sovereign (4/46)
    2858 The Essex Regiment (6/36)   61632 Royal Sovereign (10/58 — as Class B2, on withdrawal of 61671)
    2859 East Anglian (9/37 — when streamlined)
5 Locomotives marked* were reconstructed to Class B2 by Edward Thompson, including the fitting of a diagram 100A,
  B1 type boiler. Full details are to be found on pages 68 et seq.

Although the first locomotive was not delivered until November 1928, the remaining nine were recorded as being accepted in December that year, within the contract's stated delivery period. The next dozen, Nos 2810–2821 were ordered from Darlington Works just before the NBL-built locomotives arrived, and were intended to operate in the LNER Southern Area. The boilers for these locomotives, and a further 15 ordered from Darlington in 1929 were supplied by Armstrong Whitworth. In that same year, NBL asked the LNER for another order for B17s, but this was turned down, and the company preferred to acquire the next 40 locomotives from its own works at Darlington. The final batch of eleven was ordered from Robert Stephenson & Co in 1936, bringing the total in the class to 73.

Tenders

Two types of tender were paired with the Sandringhams, both of which were six-wheeled, but which brought contrasting appearances to the locomotives. The first design was very similar to the former GER type of tender, quite short, holding only 3,700 gallons of water, and was attached to Nos 2800–2847. Between the frames of the tenders paired with Nos 2800–2815 a Westinghouse brake cylinder and air reservoir were carried, since these locomotives were air braked. On the next series, Nos 2816–2842, steam brakes and cylinders were fitted, while the last five of these short wheelbase tenders were vacuum fitted. On Nos 2843–2847 the vacuum reservoir was removed from between the frames, and carried on the tank top at the back of the tender behind the coal space.

The GER type tender weighed 18 tons 19 cwt when empty, with an equally divided wheelbase of only 12 ft 0 in, and 4 ft 1 in diameter wheels. Weights carried on the tender axles ranged from 11 tons 10 cwt 2 qtrs at the front, through 13 tons 9 cwt 2 qtrs on the middle, to 14 tons 8 cwt on the rear axle. These short tenders carried only four tons of coal. A number of operational difficulties presented themselves with the design – footplate crews were not initially provided with anywhere to store food, while the narrow width of the bunker resulted in coal being spilled over the sides. Extensions of various forms were made to the inner coal guards, some on both sides, some on the right-hand side only, where the fire irons were kept.

When the Football Club series of Sandringhams began to appear from No 2848 onwards in 1936, they were paired with LNER standard tenders. The new series was put to work in the Southern Area on former GCR metals where the restrictions on both length and weight were nowhere near so demanding as they had been on former GER territory. The new tenders were 4,200 gallons, $7^{1}/_{2}$ tons capacity; commonly referred to as Group Standard, they weighed-in at $52^{1}/_{2}$ tons when fully loaded. The wheelbase of 13 ft 6 in was not divided equally, with a distance of 7 ft 3 in from leading to centre axle, and 6 ft 3 in to the trailing axle, while the wheel diameter of 3 ft 9 in was some 4 in less than on the old GER type.

The 4,200-gallon standard tender had originally appeared in 1924 with the K3 class 2-6-0s, and the Sandringham versions were all vacuum braked, with the reservoir carried at the rear on the tank top. The straight sides to the tender enclosed the self-trimming bunker which on Nos 2848–2861 was surrounded by separate coping sheets rivetted to the tender sides. The traditional method of rivetted construction was not used for the final batch of tenders for Nos 2862–2872, which were all welded with the coping plates integral with the tender sides. Fire irons were carried on these Group Standard tenders on the right-hand side, with two lockers for the footplate crews fitted high up on the tender front above the coal space opening.

Three types of brake gear were fitted to the B17 tenders: Westinghouse on Nos 2800–2815, steam brakes on Nos 2816–2842, and vacuum brakes on Nos 2843–2872. A number of minor alterations were made to these standard tenders during their working lives. Under British Railways ownership the tender attached to No 61665 was removed when the locomotive was withdrawn, and converted into a water carrier in 1960.

THE THOMPSON REBUILDS – CLASS B2

The Gresley era on the LNER had been characterised by a lack of standardisation in some aspects of motive power design and construction, with the company's policy dictated by stringent financial considerations. Gresley was a firm believer in three-cylinder propulsion, which doubtless led to many angry words from maintenance staff working between the frames. Almost at a stroke Edward

Thompson's succession as chief mechanical engineer gave emphasis to a much more rigorous standardisation policy, and the era of the general-purpose locomotive.

Thompson produced the two-cylinder Class B1 4-6-0 for new construction, and retained only the former NER Class S3 (LNER B16) and GER Class S69 (LNER B12) 4-6-0s, the latter because of its light weight, while the B17 and B7 4-6-0s were scheduled for rebuilding into two-cylinder machines, with a much simpler mechanical layout. The proposal for rebuilding the B17s was put forward in 1941, and led to a locomotive that was much nearer in axle loading to the design for a B12 replacement, which had proved so troublesome back in 1928. The first proposal had typical Gresley lines, with a boiler pressed to 220 lb sq/in, and a tractive effort of 24,310 lb. Thompson planned to retain the 6 ft 8in coupled wheels of the original B17, but with outside cylinders only, 20 in diameter by 26 in stroke, and 10 in diameter piston valves.

With the new B1 4-6-0s being constructed in some numbers, there was no urgency with the proposed rebuilds of the Sandringham class. In fact it was not until the autumn of 1944 that orders were placed, and in November that year No 2871 *Manchester City*, became the first rebuilt B17. This locomotive was chosen when it was due to enter Darlington Works for a heavy repair, following damage to its motion, after breaking the middle connecting rod. No 2871 was paired with an all-welded 4,200-gallon standard tender. It emerged from the works in 1945 in company with No 2816, both rebuilt as B2 class 4-6-0s.

The former GCR Robinson Class 1 locomotives, LNER Class B2, were reclassified B19.

Construction
The boilers fitted to the rebuilds were type 100A, as installed on the new B1 class 4-6-0 for mixed-traffic duties, and were a development of the Gresley design used on the Sandringhams. The working pressure was raised to 225 lb sq in and while the B2s received new cylinders and other alterations, since the B17s were also having type 100A boilers with their higher pressures fitted, the advantage of the rebuilt locomotives was questionable. Later extensive tests in comparison with

LNER Class B2 (rebuilt Class B17) 4-6-0 – Leading details

Designer	Edward Thompson		Working pressure	225 lb/sq in
Running – LNER	2803/2814–17/32/39/44/71		Grate area	27.9 sq ft
– BR	61603/7/14–17/32/39/44/71		Cylinders	
Wheel diameter			– number	2
– coupled	6 ft 8 in		– dimensions	20 in × 26 in
– bogie	3 ft 2 in			
Wheelbase	28 ft 2 in		Tractive effort	24,863 lb
			Adhesive weight	54 tons 14 cwt
Max. coupled axle load	18 tons 12 cwt			
			Fuel capacities – coal	7½ tons (with Group Standard tender)
Boiler			– water	4,200 gallons (with Group Standard tender)
– type	100A			
– diameter	5 ft 6 in			
– length	13 ft 11⅞ in			
– tubes: small	141 × 2 in o/d		Weights in working order	Tons Cwt
– tubes: large	24 × 5¼ in o/d		– locomotive	73 10
Heating surface			– tender	* *
– tubes	1493.0 sq ft			
– firebox	168.0 sq ft		– Total	* *
– Total	1661.0 sq ft			

* The final weights varied from 44 tons 2 cwt to 52 tons 0 cwt for tenders depending on whether LNER, ex-GER, or the ex-P1 class tenders were used

Superheater	344.0 sq ft
Superheater elements	24 × 1.244 in o/d

B17s and a B17 with the diagram 100A boiler demonstrated no advantage for the rebuilt locomotives. In addition to being more restricted in operation than the mixed-traffic B1, the B2 was reportedly an inferior performer to B17 No 1622 *Alnwick Castle*.

The mechanical changes included in the B2s involved no alterations to the main frames, except for the positioning of the cutouts for the bogies, which were moved a few inches further forward. The bogie was similar to that fitted to the B1, but with 12-spoke wheels, except that on No 61639, which had only 10-spoke wheels. New frame plates were fitted to all except No 2871, which continued to record frequent fractures. The cylinders, which were also the same as the B1 design, 20 in × 26 in, drove onto the middle axle, certain B17 motion parts were retained, and the maximum travel of the 10 in piston valves was $6^{21}/_{32}$ in. The two Wakefield mechanical lubricators supplied oil to both cylinders and axleboxes.

At the front end, the smokebox was the same diameter as that fitted to the B1 class. Originally a B17 type smokebox door was provided, later replaced by a door with more pronounced dishing, and at 4 ft 9 in slightly larger in diameter. Certain modifications were required inside the smokebox due to the higher pitch of the boiler, which meant a

shorter petticoat pipe fitted to the chimney. The firebox incorporated Gresley's drop grate as before, but no hopper ashpan, while the cab floor was level throughout, presenting none of the earlier difficulties of the B17s.

Thompson's rebuilding included a number of minor variations, with steam brakes for the locomotive as standard, and vacuum brake equipment for the train. The provision of steam brake cylinders under the cab instead of well forward between the frames as on some B17s, required the use of shoes acting on the front of the wheels, instead of the rear.

Tenders attached to the B2 class were subject to considerable variation. Of the ten locomotives converted most were paired with the former NE design, two came from former P1 class 2-8-2 locomotives, and No 2871 retained its group standard design. The ex-NER C7 Atlantics had provided complete tenders, although it was considered at one time that some existing partly completed tank bodies could be paired with the tender frames only. Some alterations were made to the second-hand tenders that were used, varying from the installation of Westinghouse equipment on the NER tenders fitted to Nos 1607 and 61644, to the shearing of over 1 ft from the leading edge of the former P1 class tender frames.

B2 Class 4-6-0 Building and Withdrawal

Running No	Rebuilt No	BR No	Date rebuilt	Withdrawn	Where scrapped
2803	1603	61603	10/1946	9/1958	Doncaster
2807	1607	61607	5/1947	12/1959	Stratford
2814	2814*	61614	11/1946	6/1959	Stratford
2815	2815*	61615	4/1946	2/1959	Stratford
2816	2816*	61616	11/1945	9/1959	Stratford
2817	1617	61617	12/1946	8/1958	Stratford
2832	1632	61632	7/1946	2/1959	Doncaster
2839	2839*	61639	1/1946	5/1959	Stratford
2844	61644	61644	3/1949	11/1959	Stratford
2871	2871*	61671	8/1945	9/1958	Stratford

N.B.:
* These locomotives were renumbered in their rebuilt form by the LNER between February 1946 and December 1946, carrying running numbers 1614, 1615, 1616, 1639 and 1671 respectively.

Although originally twenty B17s were authorised for rebuilding, only ten were so treated, nine by the LNER, and one coming out in March 1949 as BR No 61644. All retained their original names after rebuilding.

Sandringham and Rebuilt Sandringham Class 4-6-0 Operations
When the original B17 class was introduced in 1928, the North British built locomotives were run-in on stopping trains between Glasgow and Edinburgh. Their first appearance south of the border on former GER metals took place in January 1929, when they were put to work on express services on the Cambridge and Colchester lines. Despite early misgivings and a derailment on the Continental Boat Train, the Sandringhams became quite popular in Essex, but the Stratford and Ipswich locomotives were less cordially acclaimed, although they did take over the boat trains from Ipswich to Manchester Central from B12s. Harwich based locomotives were also put to work on prestigious trains between Parkeston Quay and Liverpool Street. Subsequent batches of B17s from Darlington Works went to Stratford, Ipswich and Colchester, while Doncaster received three in 1931. Among the cross-country type workings in the hands of Sandringhams was Doncaster to York and Hull, and back to Doncaster with fish trains – a complete contrast from the express passenger workings in East Anglia. By 1933, Norwich had received B17s, while three were sent to Gorton in Manchester, for working over ex-GCR metals. During the early 1930s, more B17s went to East Anglian depots, while additional locomotives at Gorton gave rise to consideration of replacement of GCR types on main line expresses.

There was apparently a certain reluctance by footplatemen in some locations to accept the new locomotives, due to the popularity of

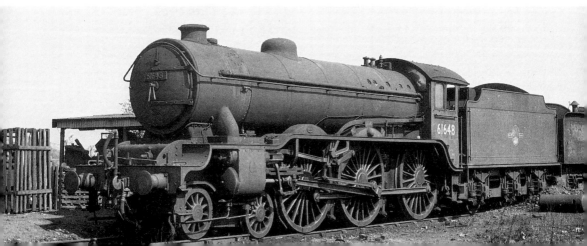

B17 No 61648 on the dump at Doncaster, awaiting scrapping in April 1959, minus nameplates. As *Arsenal*, this was one of the Football Club series, introduced with the larger 4,200-gallon tender for GCR lines. *Roger Shenton*

the former GCR Director class 4-4-0s. The increase in B17 numbers had spread to Neasden by the mid-1930s, working the Manchester-bound trains alongside the 4-4-0s. In East Anglia, while the B17s appeared on boat trains, including the Antwerp Continental and Flushing Continental, the prestigious Hook Continental was hauled almost exclusively by B12/3s. Train weights on the GE Section during the 1930s were frequently heavy, with London-bound trains from Ipswich reaching up to more than 450 tons, behind a B17.

In 1936/1937, the Football Club series emerged, and were shared between Leicester, Gorton and Neasden, dominating passenger work from Leicester in particular. Although the B17s were regarded as fast running locomotives, they were also notoriously bad riding. The use of B17s on the GE section with large tenders was limited, but the two streamlined members of the class, Nos 2859 and 2870, worked the East Anglian from September 1937. These were the first B17s with larger tenders to work regularly over this route. After World War II the East Anglian was reinstated, but with the new B1 class at its head.

On the former Great Central lines B17s were worked very hard and their characteristic bad riding showed up in problems of excessive wear in the trailing axleboxes. In 1938 Pacifics and V2s were drafted in as temporary replacements, although four years later the Pacifics were transferred back to their home areas. The same problems of axlebox wear were evident in the East Anglian locomotives, when after the war the B17s working the King's Cross and Liverpool Street trains needed intensive treatment. The generally run-down condition of the class following the exertions of the war years with very heavily loaded trains resulted from arrears of maintenance. Extensive work was required, including re-metalling of the trailing axleboxes, among many other tasks amounting to an intermediate works repair.

Sixty-four B17s remained at the end of 1947 for handing over to British Railways, plus the nine Class B2 locomotives rebuilt with the diagram 100A boiler. The latter were all based in East Anglia, with No 1671 *Royal Sovereign* becoming the official Royal Train engine. Out of the seven depots which had an allocation of B17s, twenty-four were stabled at Cambridge; the distribution remained relatively unchanged until the arrival of Britannia 4-6-2s in the 1950s.

Testing

Several series of tests were carried out with the Gresley 4-6-0s, ranging from the draughting experiments of 1929/30, through a series of trials in 1936/37 with No 2861 and the counter-pressure locomotive, to extensive dynamometer tests in comparison with a B2 in 1947. The 1937 tests involved the dynamometer car, to determine the new 4-6-0's performance over a range of speeds and cut-off positions. The testing involved using former NER Class B13 No 761 as a counter-pressure locomotive running in reverse gear, in a method of testing which Gresley wanted to detail to the press. The proposal to test a B17 was first put forward in November 1936, but due to the need to carry out a number of repairs the preparation of No 2861 took longer than anticipated, and testing began early in December 1936. The train was worked over the East Coast main line between Darlington and York, with a break over Christmas, and following cleaning of the locomotive, tests were resumed early in 1937. The report, which proved inconclusive and of little consequence in the development of the B17, was published in February 1937.

Later in its career, following an initial trial in August 1945 against a new Class L1 2-6-4T and B17 No 2827, Class B2 No 2871 took part in a series of dynamometer car trials in 1947. These tests, involving a B17 with a boiler pressed to 180 lb/sq in and the rebuilt version with Thompson's diagram 100A boiler pressed to 225 lb/sq in showed a definite advantage in favour of the B2. The comparative trials of 1947 between the two-cylinder and three-cylinder versions, demonstrated a marginal superiority for the original type B17 with the diagram 100A boiler. The diagram 100A boiler on No 1622 was set to 200 lb/sq in for the trials. The 1947 report states that the fitting of the new boiler to the B17/1 showed an economy of around 6% in coal and water consumption, and that it was on average 10% more powerful. These improvements were sufficient for the LNER to cancel further B17 to B2 conversions, and fit more diagram 100A boilers pressed to 225 lb/sq in to the original Sandringhams instead.

CHAPTER SIX

THOMPSON'S B1 4-6-0 –
A MAID OF ALL WORK

Edward Thompson succeeded Sir Nigel Gresley as chief mechanical engineer of the LNER when the latter died in office in 1941. Gresley's locomotive designs had engendered great public admiration, and had demonstrated the capability of the company's engineering practices, given the necessary service from maintenance staff. However, among the drawbacks of the pre-war philosophy was the complexity of the designs, a problem area emphasised with the shortage of skilled staff during World War II. Following the end of hostilities a very different economic picture demanded a fairly drastic change in the company's locomotive policy.

As a result of Gresley's ideas, the LNER had a range of powerful locomotives, but in a number of cases with limited route availability. For secondary services and general mixed-traffic work, a great deal of reliance was placed on designs that had been inherited from pre-Grouping companies, albeit with modifications, and a single new 4-6-0 design that had many critics and not a few

The earlier style of LNER livery for the B1 or Antelope class is seen here on No 1018 *Gnu* at York in June 1947. *M. Joyce/Gresley Society*

design flaws. Gresley had made some gesture towards the modernisation of secondary motive power with the V4 class 2-6-2, but this was subsequently seen as out of step with Thompson's simplified, all-purpose designs – one of the first actions the new cme took was to cancel orders that had been placed for the new 2-6-2s.

Contemporary thinking during the 1940s tended to encourage the production of utility types for both freight and passenger traffic, perhaps to some extent influenced by Government restrictions on the design of new passenger locomotives. This influence by Government had been an important feature in the construction of the GWR County class 4-6-0, and even Bulleid's light Pacifics for the Southern Railway. Edward Thompson's ideas culminated in the introduction of the new B1 class 4-6-0 – sometimes known as the Antelope class – in late 1942. In the magazine *Engineering* for 15 January 1943, Thompson described the new 4-6-0 as meeting the demands of both heavy passenger and of fast freight traffics over all the systems of the British railway companies, with a minimum of restrictions.

The press information also indicated that

LNER Class B1 4-6-0 – Leading details

Designer	Edward Thompson
Built	12/1942–4/1952
Withdrawn	11/1961–4/1968*
	(*From service stock)
Length	61 ft 7⅜ in
Height	12 ft 11¹¹⁄₁₆ in
Width	8 ft 8 in
Wheel diameter	
– coupled	6 ft 2 in
– bogie	3 ft 2 in
Wheelbase	28 ft 0 in (locomotive only)
Boiler – type	100A
– length	13 ft 11⅞ in
– diameter	5 ft 6 in
– tubes: small	141 × 2 in o/d
– tubes: large	24 × 5¼ in o/d
Heating surface	
– tubes	1493.00 sq ft
– firebox	168.00 sq ft
– Total	1661.00 sq ft

Superheater	
– elements	24 × 1.244 in o/d
– heating surface	344.00 sq ft
Firebox – length	9 ft 7¼ in
– width	4 ft 0½ in
Working pressure	225 lb/sq in
Grate area	27.9 sq ft
Cylinders – number	2
– dimensions	20 in × 26 in
Tractive effort	26,878 lb
Fuel capacities – coal	7½ tons
– water	4,200 gallons

	Full		Empty	
	Tons	Cwt	Tons	Cwt
Weights				
– locomotive	71	03	64	00
– tender	52	00	25	07
– Total	123	03	89	07

	Tons	Cwt	Tons	Cwt	Tons	Cwt
Axle loads						
– coupled	17	04	17	15	17	11
– bogie	16	13				

production of B1s made use of many existing jigs, tools and patterns. By this means, production costs were kept down, and the locomotive that emerged was something of an amalgam of traditional LNER design and current economic thinking. The first indications of the impending change in philosophy appeared in 1941, when a 4-6-0 design was prepared with two cylinders and the Thompson standard No 2 diagram 100A boiler. The design incorporated a new design of bogie, and 6 ft 2 in coupled wheels, while the two outside cylinders were derived from those of the K2 class 2-6-0s, and were also described as standard. The 1941 appearance of the new two-cylinder 4-6-0 was very much in the Gresley mould, with a B17 type cab, running boards with coupled wheel splashers, and a boiler pressed to 200 lb/sq in. By the second attempt in 1942 some of the details had been altered and simplified, perhaps the most noticeable of these being the absence of splashers, and higher pitched running boards. Boiler pressure was raised, becoming standardised at 225 lb/sq in.

Most of the basic design of the new standard 4-6-0 was completed by the middle of 1942, although a general arrangement drawing had been produced the previous August for the type 100A boilers. In fact, ten were ordered from Doncaster Works at the beginning of October 1941, followed by an order for ten more class B locomotive boilers in November 1941. In August 1942 an order for ten of the new B class locomotives was placed on Darlington Works, and detail drawings began to be prepared and issued. While some of the details had been taken directly from the Sandringham class, the chief draughtsman and Doncaster proposed Class B10 for the new machines. However, Thompson had other ideas, and for a time the Antelope class was simply B type until soon after the first locomotives had taken to the rails, Thompson changed his own ideas on the LNER's classification scheme, and they became Class B1. To enable the new locomotives to be classified thus the former GCR 4-6-0s became Class B18, just as the same company's B2s became Class B19 when the rebuilt Sandringhams emerged as Class B2 in the late 1940s. While the first of the new class appeared in December 1942, delivery of B1s from Darlington Works was protracted, with the final locomotive of the first order coming out in June 1944.

At the time the first locomotive appeared, the South African prime minister, Field Marshal Smuts was visiting this country, and it was decided to name No 8301 *Springbok*, with a proposal to name other locomotives after species of antelope. Although the locomotives were known as either B1 or Antelope class, the former description was more popular, while unofficially, they were nicknamed 'Bongos.' Naming was not consistent, and with more than 400 locomotives to contend with it is not surprising to find only 40 'Antelope' titles! Another 19 of the class carried names, with No 61379 *Mayflower* named in 1951 to symbolise the ties between Boston (Lincolnshire), and Boston (USA). A total of 410 was built, 290 by the North British Locomotive Co, 70 by Gorton and Darlington Works, and the remaining 50 by Vulcan Foundry.

Boiler Design
Edward Thompson's B1 design was fitted with the B17 type boiler, pressed to work at 225 lb/sq in, and designated type 100A, although the design included a number of other alterations. Built from two rings, the barrel plates were $^{11}/_{16}$ in thick – greater than the diagram 100 type fitted to the Sandringhams – because of the increase in working pressure. At an overall diameter of 5ft 6in, the rear ring was $1^3/_8$ in larger than the front ring. The smokebox tubeplates were either $^3/_4$ in or $^7/_8$ in thick. Originally, there were 143 2 in diameter tubes, but from November 1945 the number was reduced to 141, with a reduction in total heating surface to 2,033 from 2,048 sq ft.

Within the twenty-four $5^1/_4$ in diameter superheater flues $1^1/_2$ in o/d elements were installed, with a thickness of 10 swg on locomotives built up to October 1944, and 9 swg thereafter. Two different methods of attaching elements to the header were used, with two different types of header. These two types were not interchangeable, with the earlier form of attachment found on locomotives 1000–1009. The Superheater Company elements had ball-jointed (Melesco) type ends.

Steelwork for the locomotive boilers and fireboxes was, like the B17s, sub-contracted, whether the locomotives were built at the company's own works, or provided by outside firms.

Construction of the firebox followed traditional LNER practice generally, with a steel outer casing over a copper inner firebox, with $^9/_{16}$ in thick plates used for both. The length of the box was 1 in greater than that fitted to the B17 diagram 100 boilers, and the use of Monel metal was tried on the firebox

stays of Nos 1020–1025 from new. In 1950 No 61338 also received Monel metal stays, as did Nos 61400/2–5 when first built. In 1943 it was proposed that steel fireboxes be fitted, taking advantage of reduced costs, greater strength, and easier welding, but the idea did not get beyond the drawing board.

The grates installed on B1s built up to No 1035 at Darlington, and No 1210 from North British, had the Gresley drop grate with a conventional ashpan arrangement. The drop grate was evolved by Gresley, enabling the fireman to more easily dispose of any build-up of clinker. The front portion consisted of nests of bars, as a one-piece casting which could be rotated almost to a vertical position, and controlled by means of a screw in the cab acting on rodding connected directly with the front portion of the grate. The rodding in the early B1s was outside the firebox on the right-hand side, where it projected through the running board to the cab, from which point it was covered by a metal casing. From mid-1947 onwards, full rocking grates were installed, along with hopper ashpans; only Nos 61000–61009 were never altered to bring them into line with the rest of the class. Minor experiments during the 1950s included the fitting of rear damper doors on ten locomotives, although this did not improve the steaming as had been expected, and they were subsequently removed.

The drumhead pattern smokeboxes of the North British built locomotives were all-welded, while the remainder were rivetted. Originally, the blastpipe opening was 5 in diameter, but from March 1943 this was increased to $5^1/_8$ in, following experience gained with the first of the class, No 8301. From March 1944, experiments in the fitting of self-cleaning apparatus began, the first of which consisted of the fitting of simple sheet steel baffle plates in front of the tube plate, deflecting gases below the blastpipe, under a table plate. This simple layout was improved by the addition of a wire mesh screen in front of the blastpipe, and angled forwards towards the top front of the smokebox, with the idea that incandescent particles would be caught in the mesh, and not thrown out through the chimney. This arrangement was similar to that adopted later by British Railways in the Standard series of locomotives, and was installed in new B1s from No 1190 to

No 61359. Some modifications were made, aimed at improving the apparatus' performance, altering the angle of the plates and screens, curving the cement lining in the base of the smokebox, and re-positioning the smokebox door securing crossbar. In the 1950s two other forms of spark arrester were tried, consisting of mesh cages around the blastpipe and chimney petticoat pipe, although these were not so successful. During the same period, the blastpipe diameter was reduced to 4³/₄ in in order to improve the steaming qualities of locomotives fitted with self-cleaning apparatus.

Fastening the smokebox door was the conventional two handle arrangement, although the spacing of the hinges did affect the positioning of smokebox door numberplates in BR days. The degree of dishing of the door also varied slightly, some being more pronounced than others; the 4 ft 5¹/₄ in diameter door fitted to Nos 8301–8310 was the same as that fitted to the B17s in 1938. Normally the door was 4 ft 9 in diameter.

Frames, Wheels and Motion
The locomotive main frames were constructed from 1¹/₈ in thick mild steel plate, stayed apart by fabricated stretchers, with a large rectangular hole cut in the plates behind the trailing axle to reduce the locomotive weight. The frames of the B1 were

One of the Vulcan Foundry built B1s, No 61162, waits at Newcastle-on-Tyne station on 14 June 1958. *Roger Shenton*

significantly shorter from front bufferbeam to dragbox than on many of their contemporaries, particularly comparable classes like the LMS Class 5 and Hawksworth's general-purpose two-cylinder locomotives for the GWR. There was a difference between the Darlington-built locomotives and those constructed by North British – the frames of the latter were some 2 in longer than those from Darlington Works. After World War II steel castings were once again used for frame stretchers, with a slightly altered design to take account of operating experience with Thompson's new bogie. An interesting fault presented itself in later years on locomotives with a fabricated dragbox, where injectors tended to shake loose from their mountings – the remedy was to provide stays to secure the components in position.

The axleboxes for the B1s were poorly proportioned for locomotives from which so much may have been anticipated, and at 8³/₄ in diameter by 9 in long they were the same size as those used on the small V4 class 2-6-0. Contributing to the deteriorating ride with the B1s was the reduction of reciprocating balance by 36% (intended to reduce hammer-blow) and the absence of axlebox wedges. It is interesting to reflect that the arrangement of axlebox and guides used on the B1s had been used by Gresley in 1934 in an attempt to eliminate knocking with the V1 class 2-6-2 tank engines. The deteriorating ride quality of the B1 after running between 40,000 and 70,000 miles resulted in some fairly extensive repairs to keep the axleboxes and guides in good running order.

Main suspension was by means of the traditional underhung leaf springs, built up from fifteen 5 in by $\frac{1}{2}$ in thick plates, with a span between hangers of 4 ft 0 in. In the late 1950s an extra plate was installed on the trailing axle assembly. The front end of the locomotive was supported on a new design of four-wheeled bogie, with a maximum load of 18 tons 13 cwt, using laminated springs for main suspension, and helical springs for controlling side play. The mild steel frame plates were stayed apart with fabricated stretchers, although like the main locomotive frames, steel castings were used again after the war. Changes in the design involved the use of coil springs for main suspension, and a reduction in the permitted amount of side play. These minor variations in the Thompson bogie design were largely not discernible externally, and the final modifications to earlier designs with laminated main springs involved inserting additional plates and auxiliary coil springs. These alterations were applied to the class in the early 1950s, when heavy repairs were carried out.

The two outside cylinders, 20in diameter by 26 in stroke, with their 10 in diameter piston valves above, were operated by Walschaerts valve gear, and drove onto the middle coupled axle. The pattern for casting the B1 cylinders was derived from the K2 class, but with improvements in the layout of steam passages. The whole assembly was inclined at an angle of 1 in 50 when bolted to the locomotive frames, with the valves having a maximum travel of $6^{21}/_{32}$ in in fore and back gear, and a maximum cut-off of 75%.

Wheels were cast-steel centres with tyres shrunk on and fastened, to give a diameter on tread of 6 ft 2 in for the coupled wheels, and 3 ft 2 in for the bogie wheels. The centres were based on the V2 design, and their use on the new 4-6-0 was said to be in the interests of economy, although alterations in the size and position of balance weights proved necessary. In 1954, the question of the degree of balancing of the reciprocating masses came under scrutiny as a result of the ongoing ride problems, and No 61035 was tested on the Leeds to Hull route, trailing the dynamometer car. The balance weights on the locomotive were increased for the tests, and following their completion, other locomotives were similarly altered.

The lubrication of the front end was effected by means of Wakefield No 7 mechanical lubricators, carried on the left-hand running board, and driven from the reversing link. Some minor changes were made to this arrangement, but essentially these only concerned the change in the models of lubricator installed, or the drive. Oil boxes with trimmings, and six syphon feeds were car-

Class B1 4-6-0 No 61379 *Mayflower* heads a King's Cross-Cleethorpes train at Oakleigh Park in September 1960. *Derek Cross*

ried just above the running boards, mounted on the boiler cleading at the front end on both sides. Four feed trimming boxes were mounted a little further back, fixed to the boiler, just in front of the reversing link. Oil feeds from these boxes were provided for such surfaces as axleboxes and horns.

Brake shoes were carried ahead of the coupled wheels, on single hangers, and for the locomotive were steam-operated. The 9 in diameter single-acting cylinder was mounted below the cab floor between the locomotive frames, with a vacuum-operated graduable steam brake valve controlling the brake application. The vacuum brake equipment for the train included a Gresham & Craven combination ejector.

Details

Perhaps one of the most noticeable features of the appearance of the new 4-6-0s was the absence of coupled wheel splashers, with running boards carried high over the coupled wheels. These platforms were attached to the main frame by means of cantilever brackets, and were 6 ft 5 in above rail level. At the front end, there was a marked change in footplate levels, joined by a curve, with the front frame extensions protruding over the footplate and curving down to the front buffer beam. Small steps were attached on either side of these extensions, with front footsteps giving access to the running boards on both sides of the locomotive. From front buffer beam to cab spectacle plate the footplate was 8 ft 2 in wide overall, which was increased at the sides of the cab to 8 ft 8 in, to provide footholds for the crews.

The cab reached a maximum height of 12 ft 10⁹/₁₆ in over the ventilator, and incorporated two side windows – very much LNER in appearance. A departure was made with the cab floor which was level in the new design, where traditionally Gresley locomotives and LNER practice had favoured a well-type layout. The Gresley pull-out regulator was installed originally, but this was later abandoned in favour of the type most commonly seen on former GCR classes. The regulator valve was mounted in the steam dome.

Electric lighting was to be fitted to post-war builds, supplied from alternators, and similar to the arrangements used on some A2/1 and A1/1 Pacifics. The drive from the

rear axle of the bogie was supplemented by a battery mounted in the cab, with the equipment supplied by Metropolitan Vickers. The installation of the equipment seemed to be a little haphazard, with some locomotives acquiring wiring harnesses only, and some with modified lamp irons and hollowed ends to the bogie axles, but no alternators. The alternators when they were fitted gave sufficient trouble to justify the introduction of steam-driven generators in their place. Again, though, there was some inconsistency in their fitting, and the electric lighting equipment was later removed from a number of generator-fitted locomotives, and all those with the axle-mounted alternators.

Other minor detail differences in later builds included the fitting of speedometers on the majority of B1s coming out after 1946, mounted on the left-hand side in front of the cab.

The Thompson 4-6-0s in Scotland on the former GNS section were equipped with tablet catchers for single-line working. Builders' or makers' plates when fitted varied in size, shape and position, while the LNER-built locomotives had large oval brass works plates attached to the front of the locomotive main frames.

In BR service, the rear footsteps provided a suitable location for the battery boxes for the standard AWS equipment.

Tenders

Basically tenders attached to the B1s were the LNER standard 4,200-gallon type on a six-wheeled underframe, with a wheelbase of 13 ft 6 in. There were a number of minor differences between the various batches, with slight changes to the profile of the frames, the fitting of steam or vacuum brake equipment, and depending on whether the water tanks were welded or rivetted. The majority of tenders were welded, while ten originally allocated to Nos 61350–61359 built at Darlington Works in 1949 used snaphead rivets in their construction.

The 4,200-gallon standard tender first appeared in 1924 on the K3 class 2-6-0, and was also seen behind the D49, O2, J39, V2 and B17 Sandringham classes. Disc wheels of 3ft 9in diameter were used, with axles spaced at 7ft 3in and 6ft 3in, with a weight distribution which gave a tender slightly heavier at the rear than the front. The weights were 16 tons

8 cwt on the leading axle, 17 tons on the centre, and 18 tons 12 cwt on the rear. This gave a total weight in working order of 52 tons, but on the tenders constructed by North British, the weight stated was some 6 cwt less, at 51 tons 14 cwt. The outside bearing axleboxes with overhung leaf springs used two different lengths of hanger, with the mounting points for the latter positioned mid-way between top and bottom edges of the frames.

Up to locomotive No 1040 there was a greater variety in the design of tenders, with Nos 8301–8310 receiving new tenders; Nos 1010, 1011 had tenders which had originally been paired with A2 Pacifics. Similarly in 1946/47, just as the second Darlington batch was being constructed, the works was reconditioning tenders, and two which had previously run with C7 Atlantics were attached to Nos 1038 and 1039. The tenders of Nos 1012 to 1037 had higher front plates, as the floor of a B1 was higher than other LNER classes running with this type of tender.

At the end of 1945 a new tender frame drawing was put out, which was first applied to North British locomotives from No 1040 onwards. In this arrangement, standard for more than 300 B1s, there were small side buffers, while guard irons were applied to the rear of the tender frames. These had disappeared on North East area tenders after 1938, and were not provided on Darlington-built tenders after that date. Later tenders paired with Nos 61340–61409 had straight edges to the frame ends instead of the curves as previously, and in these later builds they matched the cab footsteps.

The tanks of the welded tenders attached to Nos 1040–61349 and Nos 61360–61399 had completely flush sides, while the tenders of rivetted construction had the coping plates overlapping the outside of the tender sides. Tenders for B1s built at Gorton Works were also welded, and since there was no approved drawing showing the method for welding tanks, the practices and drawings of Vulcan Foundry were used. While the Darlington-built locomotives Nos 61350–61359 and Nos 61400–61409 had rivetted tenders, the final batch used traditional Darlington Works practice with countersunk rivets, whereas four locomotives had tenders with snap-head rivets. These latter were used by Doncaster Works and were first seen on tenders paired

with Nos 61352/3/4/7. The tanks had been earmarked for a batch of B2s, although in their turn they had been ordered during the war as part of an order for O2 class 2-8-0s, which was later cancelled. While intended for the rebuilt Sandringhams, they were altered slightly to suit the new B1 4-6-0s.

At the front of the tender, the handbrake screw was carried on the left-hand side, with water scoop control screw gear on the right. The pick-up gear was carried between the frames, just behind the centre axle, with the dome covering the top of the inlet pipe on top of the tank, behind the rear divider plate for the coal space. Vertical handrails fitted to the tender sides – one at each of the four corners – were standard for all 4,200-gallon tenders, but the B1s did not receive the horizontal handrail across the back of the tender, seen on some other locomotive types. Fire iron racks were fitted inside the left-hand coping plate, with tool boxes above the coal space doors at the front, on both sides.

Tenders Nos 4095, 4155, 4200 and 4219 were converted in the early 1950s to have self-weighing coal bunkers. These modified bunkers were supported at three points, on each side of the coal space, with alterations to the fitting of toolboxes, and an extended shovelling plate. The main graduated bar marked from 0 to 8 tons, with a second bar graduated from 0 to 10 cwts, were fixed to the top of the tank just behind the bunker. Water capacity was reduced in these modified tenders to 3,750 gallons.

These coal-weighing tenders were under the control of the motive power department and were not so extensively used as their counterparts on the LMSR, which were under the control of the chief mechanical engineer. They were paired and operated with a number of different B1s during the 1950s and 1960s. The majority of B1 tenders came into the following categories:

B1 4,200 Gallon Capacity Tenders

BR Running Nos	Tender Nos	Tank construction	Frame ends
61040–61339	4030–4329	Welded	Curved
61340–61349	4330–4339	Welded	Straight
61350–61359	4340–4349	Rivetted	Straight
61360–61399	4350–4389	Welded	Straight
61400–61409	4390–4399	Rivetted	Straight

Construction

Under the LNER scheme of things, the B1s were the company's equivalent to Stanier's Black Five 4-6-0s on the LMSR. They were LNER Southern Area load class 3, route availability code 5. The eventual total of 410 locomotives was built in nine batches, with the two largest orders for 100 and 150 being placed with the North British Locomotive Company in 1945 and 1946 respectively. These two large orders, whose placement was separated by less than six months, included the largest single contract ever awarded for motive power by the LNER, and at £14,895 for each locomotive the two orders were worth some £3,723,750.

The B1 design figured as a prominent part of the LNER's modernisation programme announced in 1945 to cover a five-year period, although the first 10 locomotives had by then been in service for almost three years. Darlington Works received the first major order for 30 locomotives in 1944, followed by the two big contracts for NBL in August 1945 and January 1946. Also in January 1946 fifty were ordered from Vulcan Foundry, but the price had risen to £15,300 each. Gorton and Darlington shared the next orders in December and November 1947, building ten each, while in September 1948, the newly-formed British Railways ordered 40 more from North British, but with the price then up to £16,190 each. Finally, the last batch of 10 was ordered from Darlington Works in February 1949.

Building the new 4-6-0s occupied $7\frac{1}{2}$ years, from December 1942 to June 1950, representing an average of more than four locomotives a month over that period. The largest orders from North British were completed in just two years, at around ten locomotives a month. In order to speed-up delivery of NBL's first order, both Gorton and Darlington Works were sub-contracted to build the diagram 100A boilers, supplying 30 and 20 each respectively.

There were no major modifications to this simple, rugged design, although the problem of deterioration and fracture of firebox plates did seem to indicate a need for some improvement in that area. At one stage in their later career in order to overcome the problem it was even proposed that the locomotives be fitted with the successful BR Standard No 3 boiler, as carried by the Class 5 73XXX locomotives. The heavier loads which this would have imposed on the coupled axles would have reduced the B1s' route availability, and pushed the maximum axle load up from 17 tons 15 cwt to 18 tons 14 cwt. While the design work was being carried out and diagrams issued in the mid-1950s the end of steam rail traction was in sight, and the plans were not pursued. Some success had been achieved in strengthening the type 100A boiler at the firebox flanges, which was a further reason for the eventual decision to drop the plans to re-boiler the B1s.

Operation and Allocations

The LNER's former GE Section was once again the first to see the introduction of a new class of locomotive, just as it had been with the Sandringhams, and as it was destined to be later with the Standard Britannia Pacifics. The first of the class, No 8301 was initially allocated to Gorton depot in Manchester from March 1943, with No 8302 going to Stratford in East London in June the same year. It is interesting to note that while the B1s were counted as secondary locomotives for operating purposes on the LNER, on the GE lines initially at least they were put to work on main line expresses. Subsequent additions to the class after the war saw similar allocations, with their roles extended on former GCR metals.

From the first large order from the North British Locomotive Co some B1s were allocated to the North Eastern and Scottish areas, but the majority were given to the Southern Area. The largest number of these were seen on former GCR metals, out of a total of 51 distributed within the Western Section of the Southern Area. But it was on the GE section that the B1s first saw success, and out of this first NBL order 20 were sent to Ipswich and Norwich, showing their paces on expresses between Liverpool Street and Norwich. The reinstatement of the East Anglian after the war saw its inaugural run behind No 1048. By the time delivery of the Vulcan Foundry order began in mid-1947, more than 50 of the new 4-6-0s were at work on the Southern Area Eastern Section. The largest allocations were at Stratford and Norwich, although by the end of 1947 Doncaster had an allocation of 19 locomotives out of the class total which then stood at 274.

On former Great Central lines No 8301

was the sole member of the class resident at Gorton until 1944, when No 8307 arrived. It was not until 1946 that B1s arrived in increasing numbers and by nationalisation allocations to 11 depots totalled just under 100. The majority were at Sheffield and Gorton, working all manner of passenger and freight trains. The new locomotives seemed popular with footplate crews, and in the final years of LNER ownership they were predominant on express passenger workings. Some time-keeping problems arose from the increased post-war loadings of trains like the Master Cutler and South Yorkshireman workings, and Gresley Pacifics found their way back to GCR metals for a time. This return lasted until the early 1950s, and B1s were confined in general to workings north of Leicester. The electrification of the Woodhead route through the Pennines in the early 1950s also eliminated another common working for B1s, between Manchester and Sheffield.

The former Great Northern main line out of King's Cross saw its first B1s in the autumn of 1946, when they began to be employed on outer suburban services. Here too they were very popular locomotives, and were frequently used in place of V2s and Pacific types. Early in British Railways days they were put to work on Cambridge expresses and Peterborough semi-fasts. During the summer excursions were frequently hauled by B1s, while the King's

Another of the famous 'Butlins Express' workings, but this time an extra train, departing from King's Cross for Skegness at 8.25am on 9 June 1962 behind B1 No 61179. *R. F. Orpwood/Gresley Society*

Cross to Skegness Butlin's specials, along with other expresses, parcels and some goods workings, were the order of the day towards the end of the 1940s. The new 4-6-0s worked throughout the former GNR system, to Leeds, Hull, Sheffield and Bradford, and in East Lincolnshire. Some stopping passenger services into King's Cross, and branch passenger trains from Grantham, along with the fish trains from Immingham – once the preserve of ex Great Central 4-6-0s – were regularly in the charge of Thompson 4-6-0s.

In the North Eastern Area and in Scotland the new 4-6-0s were used for widely varied duties. Scotland in particular saw their early work mainly on passenger duties although the former Great North of Scotland line provided other work as well. In North East England the first Darlington-built locomotive No 8301, was used on the 2.00pm Darlington to Leeds working on 23 December 1942. Although joined at Darlington by nine more of the class, all were transferred away from the North East in 1943. The first regular workings for B1s allocated to North East depots began in 1946, when Leeds (Neville Hill) took over top Newcastle passenger services; these locomotives were soon replaced by Pacifics, and the 4-6-0s were relegated to secondary work. The majority of locomotives from the bulk orders placed with contractors were allocated to Darlington, Gateshead, and York, and by 1952 the total number of B1s in the North East Area had reached 80.

While most of the class were put to work on mixed-traffic duties, Gateshead-based locomotives were employed predominantly on passenger workings. In contrast, the handful at Borough Gardens worked only goods trains. York-based B1s saw wide ranging trips, running out to both East and West coasts, the latter workings comprising freight services into Liverpool.

The GE Section, which had already proved itself to be something of a problem in regard to the design of locomotives, especially with the B17 class 4-6-0s, witnessed a tremendous increase in loadings during World War II. The earliest arrival of one of the new 4-6-0s was No 8302 in June 1943, and by all accounts was a success. East Anglian three-cylinder B17s had suffered greatly during the hostilities, and certain areas covered by the class were the subject of

the attention of the Luftwaffe. Two more B1s were allocated to Stratford by 1945, joining No 8302, with two more to Parkeston, Ipswich, and Cambridge. Again, the bulk orders placed with contractors resulted in greater numbers of B1s appearing during mid 1946, and almost immediately they took over the principal express passenger workings, from bases at Norwich and Ipswich.

Displacing the Sandringhams from front rank passenger duties, the class took over the re-instated East Anglian express in late 1946, albeit on a slightly easier schedule, but more heavily loaded. While a further allocation of B1s took place before nationalisation, these were restricted mainly to Stratford, Ipswich, Parkeston and Cambridge. By the early

1950s, more than seventy B1s were at work on the GE Section, with March added to the list of depots housing B1s.

As a whole, the class was well received and popular, although their condition and performance began to deteriorate, and during the BR period it gave rise to serious consideration for major modification. These are discussed in the final chapter, along with the detailed testing and performances the locomotives put up in various trials.

B1 Class 4-6-0 – Locomotive names

BR No	Name						
61000	Springbok	61015	Duiker	61030	Nyala	61238	Leslie Runciman
61001	Eland	61016	Inyala	61031	Reedbuck	61240	Harry Hinchliffe
61002	Impala	61017	Bushbuck	61032	Stembok	61241	Viscount Ridley
61003	Gazelle	61018	Gnu	61033	Dibatag	61242	Alexander Reith Gray
61004	Oryx	61019	Nilghai	61034	Chiru	61243	Sir Harold Mitchell
61005	Bongo	61020	Gemsbok	61035	Pronghorn	61244	Strang Steel
61006	Blackbuck	61021	Reitbok	61036	Ralph Assheton	61245	Murray of Elibank
61007	Klipspringer	61022	Sassaby	61037	Jairou	61246	Lord Balfour of Burleigh
61008	Kudu	61023	Hirola	61038	Blacktail	61247	Lord Burghley
61009	Hartebeeste	61024	Addax	61039	Steinbok	61248	Geoffrey Gibbs
61010	Wildebeeste	61025	Pallah	61040	Roedeer	61249	Fitzherbert Wright
61011	Waterbuck	61026	Ourebi	61189	Sir William Grey	61250	A. Harold Bibby
61012	Puku	61027	Madoqua	61215	William Henton Carver	61251	Oliver Bury
61013	Topi	61028	Umseke	61221	Sir Alexander Erskine-Hill	61379	Mayflower
61014	Oribi	61029	Chamois	61237	Geoffrey H. Kitson		

LNER Class B1 4-6-0 – Building and withdrawal

Running No	Date built	Works No	Withdrawn	Running No	Date built	Works No	Withdrawn
Built by Darlington Works				1068	8/46	25824	6/63
1000	12/42	–	3/62	1069	8/46	25825	8/63
1001	6/43	1912	9/63	1070	8/46	25826	8/65
1002	9/43	1916	6/67	1071	8/46	25827	2/63
1003	11/43	1920	12/65	1072	9/46	25828	5/67
1004	12/43	1922	12/63	1073	9/46	25829	9/63
1005	2/44	1925	9/62	1074	9/46	25840	9/63
1006	3/44	1927	9/63	1075	9/46	25831	9/63
1007	4/44	1928	2/64	1076	9/46	25832	9/65
1008	5/44	1931	12/66	1077	9/46	25833	5/62
1009	6/44	1934	9/62	1078	9/46	25834	10/62
1010	11/46	1990	11/65	1079	9/46	25835	6/62
1011	11/46	1991	11/62	1080	9/46	25836	3/64
1012	11/46	1992	6/67	1081	10/46	25837	6/64
1013	12/46	1993	12/66	1082	10/46	25839	12/62
1014	12/46	1994	12/66	1083	10/46	25838	9/63
1015	1/47	1995	11/62	1084	10/46	25841	6/64
1016	1/47	1996	10/65	1085	10/46	25842	11/61
1017	1/47	1997	11/66	1086	10/46	25840	12/62
1018	2/47	1998	11/65	1087	10/46	25843	12/65
1019	2/47	1999	3/67	1088	10/46	25844	9/63
1020	2/47	2000	11/62	1089	10/46	25845	4/66
1021	3/47	2001	6/67	1090	10/46	25846	9/63
1022	3/47	2002	11/66	1091	10/46	25847	9/62
1023	4/47	2003	10/65	1092	10/46	25848	2/66
1024	4/47	2004	5/66	1093	11/46	25849	7/65
1025	4/47	2005	12/62	1094	11/46	25850	6/65
1026	4/47	2006	2/66	1095	11/46	25851	12/63
1027	5/47	2007	9/62	1096	11/46	25852	9/62
1028	5/47	2008	10/62	1097	11/46	25853	1/62
1029	6/47	2009	12/66	1098	11/46	25854	7/65
1030	6/47	2010	9/67	1099	11/46	25855	9/66
1031	7/47	2011	11/64	1100	11/46	25856	11/62
1032	8/47	2012	11/66	1101	11/46	25857	12/66
1033	8/47	2013	3/63	1102	12/46	25862	4/67
1034	10/47	2014	12/64	1103	12/46	25863	7/66
1035	10/47	2015	12/66	1104	12/46	25864	4/64
1036	11/47	2016	9/62	1105	12/46	25858	12/64
1037	11/47	2017	5/64	1106	12/46	25859	11/62
1038	12/47	2018	5/64	1107	12/46	25861	8/65
1039	12/47	2019	6/65	1108	12/46	25860	12/62
				1109	12/46	25865	7/64
Built by North British Locomotive Company				1110	12/46	25866	10/65
				1111	12/46	25867	9/62
1040	4/46	25796	7/66	1112	12/46	25868	12/62
1041	4/46	25797	4/64	1113	12/46	25869	9/63
1042	5/46	25799	4/66	1114	1/47	25870	9/62
1043	5/46	25800	7/62	1115	1/47	25871	5/67
1044	5/46	25801	3/64	1116	1/47	25872	7/66
1045	5/46	25802	9/62	1117	1/47	25873	2/64
1046	5/46	25798	4/62	1118	1/47	25874	7/64
1047	6/46	25803	9/62	1119	1/47	25875	11/63
1048	6/46	25804	9/62	1120	1/47	25876	1/65
1049	6/46	25805	11/65	1121	1/47	25877	4/66
1050	6/46	25806	2/66	1122	1/47	25878	11/63
1051	6/46	25807	2/66	1123	1/47	25879	5/67
1052	6/46	25808	9/62	1124	2/47	25880	9/62
1053	6/46	25809	2/63	1125	2/47	25881	12/63
1054	6/46	25810	9/62	1126	2/47	25882	9/63
1055	7/46	25811	2/66	1127	2/47	25883	8/65
1056	7/46	25812	4/64	1128	2/47	25884	12/62
1057	7/46	25813	4/50	1129	2/47	25885	9/65
1058	7/46	25814	2/66	1130	2/47	25886	9/62
1059	7/46	25815	11/63	1131	2/47	25887	12/66
1060	8/46	25816	9/62	1132	2/47	25888	9/66
1061	8/46	25817	9/65	1133	2/47	25889	9/66
1062	8/46	25818	8/64	1134	3/47	25890	10/65
1063	8/46	25819	3/66	1135	3/47	25891	9/63
1064	8/46	25820	10/62	1136	3/47	25892	10/62
1065	8/46	25821	9/64	1137	3/47	25893	5/62
1066	8/46	25822	9/62	1138	3/47	25894	1/65
1067	8/46	25823	12/62	1139	4/47	25895	9/62

Running No	Date built	Works No	Withdrawn	Running No	Date built	Works No	Withdrawn
Built by Vulcan Foundry				1210	7/47	26111	2/66
1140	4/47	5498	12/66	1211	7/47	26112	11/62
1141	4/47	5499	7/65	1212	7/47	26113	11/64
1142	4/47	5500	9/63	1213	7/47	26114	4/64
1143	4/47	5501	2/64	1214	7/47	26115	5/65
1144	4/47	5502	4/64	1215	7/47	26116	3/65
1145	4/47	5503	1/66	1216	7/47	26117	1/67
1146	4/47	5504	3/64	1217	8/47	26118	3/62
1147	4/47	5505	12/65	1218	8/47	26119	7/65
1148	4/47	5506	9/66	1219	8/47	26120	6/64
1149	4/47	5507	9/62	1220	8/47	26121	10/65
1150	4/47	5508	9/62	1221	8/47	26122	3/65
1151	4/47	5509	9/62	1222	8/47	26123	1/62
1152	5/47	5510	4/64	1223	8/47	26124	1/66
1153	5/47	5511	1/65	1224	8/47	26125	7/66
1154	5/47	5512	9/62	1225	8/47	26126	6/65
1155	5/47	5513	3/64	1226	8/47	26127	9/62
1156	5/47	5514	11/63	1227	8/47	26128	9/63
1157	5/47	5515	8/65	1228	9/47	26129	9/62
1158	5/47	5516	4/66	1229	9/47	26130	6/64
1159	5/47	5517	9/63	1230	9/47	26131	11/62
1160	5/47	5518	9/63	1231	9/47	26132	7/62
1161	5/47	5519	12/66	1232	9/47	26133	2/66
1162	5/47	5520	12/64	1233	9/47	26134	11/63
1163	5/47	5521	9/62	1234	9/47	26135	8/62
1164	5/47	5522	9/62	1235	9/47	26136	9/62
1165	5/47	5523	11/64	1236	9/47	26137	9/62
1166	5/47	5524	9/62	1237	9/47	26138	12/66
1167	5/47	5525	12/64	1238	9/47	26139	2/67
1168	6/47	5526	10/65	1239	10/47	26140	8/62
1169	6/47	5527	12/63	1240	10/47	26141	12/66
1170	6/47	5528	7/62	1241	10/47	26142	12/62
1171	6/47	5529	9/62	1242	10/47	26143	7/64
1172	6/47	5530	12/65	1243	10/47	26144	5/64
1173	6/47	5531	1/67	1244	10/47	26145	10/65
1174	6/47	5532	12/63	1245	10/47	26146	7/65
1175	6/47	5533	12/63	1246	10/47	26147	12/62
1176	6/47	5534	11/65	1247	10/47	26148	6/62
1177	6/47	5535	9/63	1248	10/47	26149	11/65
1178	6/47	5536	2/64	1249	10/47	26150	6/64
1179	6/47	5537	1/65	1250	10/47	26151	4/66
1180	6/47	5538	5/67	1251	11/47	26152	4/64
1181	7/47	5539	11/63	1252	11/47	26153	11/63
1182	7/47	5540	9/62	1253	11/47	26154	9/62
1183	7/47	5541	7/62	1254	11/47	26155	9/62
1184	7/47	5542	12/62	1255	11/47	26156	6/67
1185	7/47	5543	10/64	1256	11/47	26157	11/65
1186	7/47	5544	11/62	1257	11/47	26158	10/65
1187	7/47	5545	9/62	1258	11/47	26159	1/64
1188	7/47	5546	11/65	1259	11/47	26160	8/65
1189	8/47	5547	5/67	1260	11/47	26161	12/62
				1261	11/47	26162	9/66
				1262	12/47	26163	4/67
Built by North British Locomotive Company				1263	12/47	26164	12/66
				1264	12/47	26165	11/65
1190	5/47	26091	6/65	1265	12/47	26166	2/62
1191	5/47	26092	8/65	1266	12/47	26167	9/62
1192	5/47	26093	10/62	1267	12/47	26168	12/62
1193	5/47	26094	9/62	1268	12/47	26169	12/64
1194	5/47	26095	8/65	1269	12/47	26170	12/63
1195	5/47	26096	11/65	1270	12/47	26171	9/63
1196	5/47	26097	9/65	1271	12/47	26172	7/62
1197	5/47	26098	6/64	1272	12/47	26173	1/65
1198	6/47	26099	4/65	1273	1/48	26174	5/63
1199	6/47	26100	1/67	1274	1/48	26175	11/64
1200	6/47	26101	12/62	1275	1/48	26176	10/65
1201	6/47	26102	1/62	1276	1/48	26177	6/65
1202	6/47	26103	9/62	1277	1/48	26178	6/64
1203	6/47	26104	7/62	1278	1/48	26179	4/67
1204	6/47	26105	11/63	1279	1/48	26180	9/63
1205	6/47	26106	11/63	1280	1/48	26181	9/62
1206	7/47	26107	9/62	1281	1/48	26182	2/66
1207	6/47	26108	12/63	1282	1/48	26183	9/62
1208	7/47	26109	9/65	1283	2/48	26184	9/62
1209	7/47	26110	9/62	1284	2/48	26185	9/62

Running No	Date built	Works No	Withdrawn	Running No	Date built	Works No	Withdrawn
1285	2/48	26186	12/65	61355	9/49	2077	6/64
1286	2/48	26187	9/62	61356	9/49	2078	7/64
1287	2/48	26188	9/62	61357	10/49	2079	6/65
E1288	2/48	26189	1/64	61358	10/49	2080	12/63
E1289	2/48	26190	6/67	61359	10/49	2081	12/63
E1290	2/48	26191	3/62				

Built by North British Locomotive Company

Running No	Date built	Works No	Withdrawn	Running No	Date built	Works No	Withdrawn
E1291	2/48	26192	5/65	61360	3/50	26819	4/66
E1292	2/48	26193	9/65	61361	3/50	26820	12/65
E1293	2/48	26194	8/66	61362	3/50	26821	9/62
E1294	3/48	26195	11/64	61363	4/50	26822	9/62
E1295	3/48	26196	11/62	61364	4/50	26823	9/62
E1296	3/48	26197	11/62	61365	4/50	26824	7/65
E1297	3/48	26198	11/62	61366	4/50	26825	12/62
E1298	3/48	26199	6/62	61367	4/50	26826	8/65
E1299	3/48	26200	7/65	61368	4/50	26827	1/62
E1300	3/48	26201	11/63	61369	5/50	26828	12/63
E1301	3/48	26202	9/62	61370	10/50	26829	7/65
E1302	3/48	26203	4/66	61371	10/50	26830	9/62
E1303	3/48	26204	11/66	61372	12/50	26841	6/65
61304	3/48	26205	10/65	61373	12/50	26842	9/62
61305	4/48	26206	10/63	61374	2/51	26833	9/63
61306	4/48	26207	9/67	61375	2/51	26834	11/63
61307	4/48	26208	11/66	61376	4/51	26835	2/62
61308	4/48	26209	11/66	61377	5/51	26836	9/62
61309	4/48	26210	1/67	61378	5/51	26837	11/63
61310	4/48	26211	4/65	61379	6/51	26838	8/62
61311	4/48	26212	9/62	61380	8/51	26839	3/62
61312	4/48	26213	3/64	61381	9/51	26840	11/62
61313	4/48	26214	11/65	61382	9/51	26831	12/64
61314	4/48	26215	12/63	61383	10/51	26832	1/63
61315	4/48	26216	2/66	61384	10/51	26843	1/66
61316	5/48	26217	12/62	61385	10/51	26844	10/65
61317	5/48	26218	9/62	61386	10/51	26845	12/66
61318	5/48	26219	9/63	61387	11/51	26846	10/65
61319	5/48	26220	12/66	61388	11/51	26847	6/67
61320	5/48	26221	8/65	61389	11/51	26848	11/65
61321	5/48	26222	8/64	61390	12/51	26849	2/66
61322	5/48	26223	2/66	61391	12/51	26850	9/62
61323	5/48	26224	11/63	61392	12/51	26851	6/65
61324	6/48	26225	10/65	61393	1/52	26852	9/63
61325	6/48	26226	9/63	61394	1/52	26853	11/65
61326	6/48	26227	3/66	61395	2/52	26854	10/62
61327	6/48	26228	2/65	61396	2/52	26855	9/65
61328	6/48	26229	9/63	61397	3/52	26856	6/65
61329	6/48	26230	4/66	61398	3/52	26857	11/64
61330	6/48	26231	11/66	61399	4/52	26858	9/63
61331	6/48	26232	9/63				
61332	6/48	26233	12/62				
61333	7/48	26234	12/62				
61334	7/48	26235	12/63	Built by Darlington Works			
61335	7/48	26236	9/62				
61336	8/48	26237	9/63	61400	3/50	2102	12/64
61337	8/48	26238	9/67	61401	4/50	2103	4/64
61338	8/48	26239	1/65	61402	4/50	2104	6/64
61339	9/48	26240	11/62	61403	4/50	2105	7/66
				61404	5/50	2106	11/65
Built by Gorton Works				61405	5/50	2107	9/62
				61406	5/50	2108	4/66
61340	11/48	998	4/67	61407	6/50	2109	4/67
61341	12/48	999	12/63	61408	6/50	2110	12/62
61342	1/49	1000	12/66	61409	6/50	2111	9/63
61343	2/49	1001	3/66				
61344	3/49	1002	9/66				
61345	4/49	1003	7/66				

Notes:

1. Nos 1000–1009 carried running numbers 8301–8310 when built prior to the LNER 1946 re-numbering scheme.

2. Locomotives were re-numbered in the British Railways scheme by the addition of 60,000 to the numbers shown.

3. Locomotives from No 61304 upwards carried their British Railways numbers when built. Nos 1288 to 1304 came out with the interim 'E' prefix to their erstwhile LNER numbers.

4. See separate table for details of named locomotives.

Running No	Date built	Works No	Withdrawn
61346	4/49	1004	6/64
61347	5/49	1005	4/67
61348	6/49	1006	12/65
61349	7/49	1007	8/66

Built by Darlington Works

Running No	Date built	Works No	Withdrawn
61350	7/49	2072	11/66
61351	8/49	2073	7/64
61352	8/49	2074	10/62
61353	9/49	2075	8/65
61354	9/49	2076	4/67

CHAPTER SEVEN

THE FINAL YEARS

After 25 years of service under LNER management many of the pre-Grouping designs had been modified, altered or rebuilt in a variety of ways, resulting in numerous sub-divisions of the basic classes. By the end of 1947, the LNER possessed no fewer than 19 different classes of 4-6-0, totalling more than 500 locomotives. Although there was a substantial number of different classes, the majority of the 4-6-0s owned by the LNER came from the recently-introduced Thompson B1 class which at the time of nationalisation was still in production and not completed until April 1952.

In the first year of British Railways ownership, 593 4-6-0s came under the management of the Eastern, North Eastern, and Scottish Regions. Out of this total 340 (57%) were accounted for by the B1 design, leaving 253 other types in 18 classes and sub-classes. Three of the main constituent companies of the LNER in England were represented: Great Central, Great Eastern, and North Eastern. Despite the frequent modifications during LNER days, 98 of their 4-6-0s survived to become BR locomotives in their original guise. One of the North Eastern Railway pioneer 4-6-0s, the S class, which became Class B13 under LNER ownership, was still in departmental use. This solitary survivor of Worsdell's famous class had been used as a counter-pressure locomotive for locomotive testing for a number of years, and could be found at the Rugby Testing Station in the late 1940s. Of the remaining 97 original pre-Grouping locomotives, 28 were of Great Central origin, another 48 from the North Eastern, and 21 were Great Eastern B12s.

During the Gresley and Thompson periods substantial rebuilding of many of these 4-6-0s was carried out, with the former GER 4-6-0s subjected to the most frequent if not the most extensive alterations. The B12/3 rebuilds, of which all 50 were still in stock at nationalisation, had undergone modifications by Gresley in 1932, and by Thompson in 1943. In BR days only the unrebuilt

B12/1 series was at work on Scottish lines. Remaining classes of locomotive from the pre-Grouping era that survived to operate on the national network included 21 of Vincent Raven's mixed-traffic 4-6-0s of North Eastern origin. As Class B16 in LNER days, three sub-classes had been created: B16/1, locomotives in original form; B16/2, the seven locomotives modified by Gresley from 1937; B16/3, the 14 locomotives modified by Edward Thompson in 1944. The original Raven design, as B16/1, was still in the majority in 1948 and totalled 48 locomotives in all, although more than half of these were to disappear by 1950.

The Great Central Railway designs – both original and rebuilt varieties – offered the smallest contribution to British Railways stock, with only 39 locomotives. Having said that, it will be remembered that Robinson's 4-6-0 designs for the GCR were numerous, with no fewer than nine different classes, six of which survived just into the post-1948 era. The rebuilding accounted for two sub-classes, with BR acquiring the following ex-GCR types: B3/3, B4/3, B4/4, B5, B7/1, B7/2, B8, and B9. Of these 39 locomotives,

One of the NBL-built Class B17s, which cost some £7,280 each in 1928, is No 2805 *Burnham Thorpe*. In April 1938 this locomotive was renamed *Lincolnshire Regiment* and was later fitted with the diagram 100A boiler which proved very successful in improving the Sandringhams' performance. *L&GRP*

only 11 had been modified under LNER management, and four of James Robinson's first essays into the design of a 4-6-0 type carried British Railways running numbers.

In total then, 180 pre-Grouping designs came into the new owner's stock, with 73 locomotives of Gresley design for the LNER. Most successful were the B1 class 4-6-0s, 274 of which were handed over to BR, with more still under construction, and others as yet not even ordered. The final tally for this successful design was 409 locomotives. Despite their popularity, only two of their number were rescued in later years for restoration and preservation. While the B1s began to appear in 1942, another former GCR design was still in service, carrying the same classification B1, a type introduced by Robinson in 1903, but which had become extinct by the end of 1947.

Gresley's own design of 4-6-0, introduced back in the late 1920s, was planned and introduced originally for the GE Section, later spreading to other lines; by nationalisation 64 were in stock, in three varieties. Forty were B17/1, the original design of 1928, twenty-two were B17/4s of the Football Club series, while B17/5 covered the two fitted with the A4 type streamlined casing. A number of the Sandringham class which were still hard at work on former GER metals in the 1950s had been rebuilt, or extensively modified by Thompson just after World War II and fitted with B1 type boilers. These were designated class B2, and nine came into BR hands in 1948.

Soon after the change of ownership the B1 design was selected to take part in the Interchange Trials in the mixed-traffic category. It was perhaps an obvious choice, but not only on the grounds of numerical strength – also it was a successful synthesis of the traditional and latest LNER locomotive design policies and practices. Some aspects of this latter found their way into the design of the new range of Standard steam locomotives for British Railways.

The designs taken over from the LNER by the new administration in 1948 were as shown below.

PRE-GROUPING DESIGNS UNDER BR OWNERSHIP

North Eastern Railway
The North Eastern Railway, which gave birth to the 4-6-0 type for passenger service, was still well represented at nationalisation by the B16 series in three variations, along with one of Worsdell's original class S type. This locomotive No 1699, was based at the new Rugby Testing Station after its opening, but had been in service department use for a number of years as a counter-pressure locomotive for testing purposes. Its life was curtailed three years after nationalisation, when it was withdrawn in 1951.

The B16 series of locomotives were the old NER class S3, designed by Vincent Raven and built by the company between 1919 and the Grouping of 1923, with further examples constructed by the LNER under Gresley,

LNER 4-6-0s taken into BR stock – 1948

Class	No	Former owner	Introduced	Variations
B1	340	LNER	1942	—
B2	9	LNER	1945	B17 rebuild with 100A boiler
B3/3	1	GCR	1917/1943	Rebuild of B3/2 100A boiler
B4/3	3	GCR	1925	Rebuild by Gresley with piston valves
B4/4	1	GCR	1906	Original design, with slide valves.
B5	4	GCR	1902	Original design.
B7/1	18	GCR	1921	Original design.
B7/2	7	GCR	1923	Smaller chimney and cab fitted.
B8	2	GCR	1913	Original design.
B9	3	GCR	1906	Original design.
B12/1	21	GER	1911	Original design.
B12/3	50	GER	1932	Gresley rebuild with round top boiler.
B16/1	48	NER	1920	Original design.
B16/2	7	NER	1937	Gresley rebuild, conjugate valve gear.
B16/3	14	NER	1944	Thompson rebuild of B16/2.
B17/1	40	LNER	1928	Original design with GER tender.
B17/4	22	LNER	1936	Original design with LNER tender.
B17/5	2	LNER	1937	Streamlined casing fitted.
B13	1	NER	1906	In service use at Rugby Test Plant.

until 1924. The 69 locomotives that existed at the end of 1947 were divided among three sub-classes. Seven locomotives from the first rebuilds by Gresley had their single casting for the cylinders replaced by three separate units, and were provided with the Walschaerts/Gresley conjugate valve gear. These locomotives were still working out of York depot in 1948. The alterations included in class B16/3 comprised two sets of outside Walschaerts valve gear, and the almost standard Gresley arrangement of derived motion for the inside cylinder in three-cylinder designs.

A second rebuilding took place during the Thompson era, a period which was characterised by both experiment and by a simpler approach to locomotive design. This second rebuilding resulted in the appearance of class B16/3, of which 14 examples carried British Railways running numbers. The later rebuilds eliminated the conjugated valve gear, and installed three separate sets of Walschaerts valve motion.

For an old design, the B16s gave good service in BR days, although throughout their career they seldom ventured south of Doncaster. They were originally classed as goods engines by the NER, but were treated to the mixed-traffic lined black livery by British Railways. All 69 eventually received smokebox numberplates, but as the construction of B1s crept up on the number series for the ex-NER locomotives, the B16s took over numbers from the B9 4-6-0s which had been scrapped, in the range 61469–61478. The B16s remained largely untouched throughout the 1950s, and despite their main role as freight engines, they could often be found on expresses and summer excursions. All B16/1, B16/2, and B16/3 locomotives were allocated to York, Heaton, Hull, Scarborough, and Starbeck (Leeds) in 1957, operating in the North Eastern Region, with occasional forays onto former Great Central metals.

Between 1958 and 1960, nineteen locomotives were taken out of stock, with the 50 survivors at only three depots – Mirfield, Leeds, and York – at the beginning of the 1960s. These remaining locomotives were withdrawn as follows: 1961 26; 1962 2; 1963 9; 1964 13. All the B16s could be found at York and Hull in 1964, but had gone to the scrapyard by July that same year. The majority were cut-up at Darlington Works, with some going to scrap merchants. None was rescued for preservation.

Great Central Railway Types
The former Great Central Railway types were the first to go to the breaker's torch under the new administration, with only No 1482 *Immingham* surviving to 1950. This locomotive, as class B4, remained on former GCR territory until its withdrawal, allocated to Ardsley. All former GCR locomotives were allocated running numbers in the new scheme of numbering when it was announced in 1948, but with the rapid withdrawal of all 39 remaining locomotives in just over two years, they were not carried.

With the two LNER renumbering schemes of the 1940s, in common with all other LNER owned locomotives, the new British Railways scheme was the third such change to take place in a few years. Then for those that saw a little more service with BR further repainting was required under the new standards being devised. No 1482 carried its LNER green livery until withdrawal. It kept the original slide valves fitted to the class back in 1906 when it was built by Beyer Peacock in Manchester.

Great Eastern Railway
In 1948 seventy-four of the former GER 4-6-0s, classified B12 by the LNER, were inherited by the new organisation, twenty-one were in their original form, with small boilers, and Belpaire topped fireboxes. The remainder, as Class B12/3, were provided in 1932 with a larger boiler round-topped firebox, and long travel piston valves, along with the more cosmetic alterations to the running boards and valances. Eight of the B12/1 series, BR Nos 61504/5/7/8/11/24/6/32, had also been altered in 1943 by Thompson, and carried a smaller round-topped boiler, in comparison with the B12/3. In total, BR took over only 13 locomotives in largely original condition.

With the B12/1 and B12/3 types renumbered for a second time in 1946, numbers in the 1500 series were allocated. In 1948 these had 60,000 added, with most locomotives receiving their new numbers in 1948/49.

As Class 4P/3F 4-6-0s, these former GER types were to be found almost exclusively in East Anglia on the Eastern Region, although

B17 No 61636 *Harlaxton Manor*, seen on an up Cambridge working at Wymondley in 1953, is paired with the original GE style short wheelbase tender. Worth noting are the inner coping plates at the side of the tender coal space, to prevent spillage. *G. W. Goslin/Gresley Society*

a number of the original B12/1 series were at work on the Scottish Region. Allocated to Kittybrewster and Keith depots, the locomotives worked over former Great North of Scotland lines between Aberdeen and Elgin on passenger, freight and fish trains. With occasional visits to the West Coast of Scotland during the summer months, the B12s had been employed on such work for more than a decade by the time the railways were nationalised. Some of the class peculiarities – such as the ACFI feed water heating appar-

One of the two preserved B1s restored to LNER livery, No 1306 carrying the nameplates *Mayflower*. Although originally stabled for a number of years at Steamtown, the locomotive can now be found at Loughborough. This study was taken in September 1986. *Roger Shenton*

atus – which earned the locomotives concerned their nickname of 'Hikers' (or 'Camels' on the Eastern Region) had disappeared.

These original B12s had a route availability of 3, while the heavier and extensively modified B12/3 series were RA4. All the B12/1s were withdrawn between May 1951 and November 1954; the larger B12/3s survived for another seven years before the last, No 61572, was withdrawn in 1961.

In 1948, a number of the B12/3s allocated to East Anglian depots were transferred to Yarmouth Beach, and South Lynn depots, for working over the former M & GN line. In 1954/55 some of the B12s allocated to these duties were supplanted by the Ivatt Class 4MT 2-6-0s. The majority of B12/3s were allocated to Stratford in 1950, but by the middle of the decade some re-allocation saw a more balanced distribution of the 42 that remained. In 1957 for instance, employed mainly on local passenger and freight work-

ings on the GE Section, most B12s were found at Norwich, Ipswich, and Stratford, 26 in all. The former Midland Railway shed at Peterborough (Spital Bridge), was home to five B12s working out to Northampton. The class began working further afield in 1949, when locomotives allocated to Grantham worked up to Lincoln, and on occasions took excursion trains to York and Scarborough. In the mid-1950s, six B12s were located at Grantham for these duties.

At the end of 1958, a further seventeen B12/3s had been scrapped, all at Stratford Works, leaving only 25 on the Eastern Region books. At Yarmouth Beach, three of the class were still working over the M&GN lines, along with the LMS 2-6-0s, while the remaining 22 were located at Cambridge, Norwich and Ipswich. At the end of the 1950s dieselisation – the new multiple units and Brush Type 2 diesels which were coming into regular Eastern Region service – was making inroads into former B12 duties. However, the remaining members of the class were still working regularly over the GE Section into Liverpool Street station on passenger services and parcels trains. In 1959, only 16 were left, all stabled at Cambridge and Norwich.

Last of the B12s was No 61572, at Norwich from 1959. As the sole survivor of the class, it was maintained in good order, for regular service on passenger and parcels turns, and on rail tours until its withdrawal from service in September 1961. This locomotive was undergoing restoration on the North Norfolk Railway in 1988.

GRESLEY'S SANDRINGHAM CLASS

Sixty-four B17s and nine B2s were taken over by BR in 1948, and were renumbered by the addition of 60,000 to their post-1946 running numbers from late 1948. In 1950, the remaining Sandringhams were allocated to nine Eastern Region depots two of which, Colwick and Woodford Halse on former GCR lines, were later transferred to London Midland Region control. The largest number, 17 in all, were allocated to Cambridge, and these were all B17/1s with the GER style tender. Stratford housed another 12, with 10 each at Ipswich and March. In each case, the majority were B17/1, but re-allocations gave rise to the arrival of B17/4s with the LNER standard tender. The two

streamlined locomotives, Nos 61659 and 61670 were stabled at Norwich, with the remaining members of the class found at Colchester, Norwich, and Yarmouth (South Town). In the same year, only eight of the class which had been rebuilt as B2s were left, six at Colchester, two at Cambridge.

No 61661 *Sheffield Wednesday* was one of the first locomotives from all four main line railways to sport any sign of new ownership. Early in 1948, the newly-formed British Railways was experimenting with liveries, and No 61661 appeared in bright green livery lined out in yellow, with the words 'British Railways' on the tender. Later in 1948, the lining was changed to red, yellow and grey. No 61665 was painted in LNER apple green livery, with different lining and the new owner's name on the tender. Both were sent around the country as part of a public relations exercise designed to obtain public reaction to the new colour schemes. In addition No 61661 was allocated a rake of Gresley coaches to haul in its wake, repainted in GWR style chocolate-and-cream livery – it is hard to imagine a more unusual colour scheme!

As BR locomotives B17s were classified 4P, with a route availability code of 5, and in 1951, all were stationed at depots on the GE Section.

The Sandringhams had all gravitated to East Anglia, and been replaced by the V2 class 2-6-2s on former GCR lines, but on the GE Section they were faced with stiff competition from B1 and Standard Britannia class Pacifics in BR days. Insofar as the Britannias were concerned, B17s were called on to deputise for the Pacifics on occasions, although the timings were such that there were problems arising from incidents of overheating of the middle big-end. Some depots, such as Norwich, preferred the Thompson B1 to the three-cylinder B17 for many duties. Most of the class spent their BR careers on the easy profiles of the routes through the fenlands.

Following the end of World War II it was some time before international traffic rose to a level sufficient to justify re-introduction of boat trains. These services from Parkeston to Manchester were worked by Parkeston B17s as far as Lincoln, but from 1950 the train was worked up to Sheffield by a B17, usually from March depot. Some Sandringhams did sterling work on passenger duties, especially

An attractive and typical view of a B16 in BR days, as No 61446 hauls a rake of coaches in the carmine and cream livery of the 1950s. *G. W. Sharpe*

on the East Suffolk lines, and took over regular working of the recently instituted East-erling service from 1952. The Yarmouth locomotives which provided the motive power for this working replaced the B1s which had been previously employed on the train from its inauguration.

The GE Section of the former LNER lines seemed always to have been the centre of controversy in terms of motive power, and in BR days the story continued. The introduction of the first Standard designs of steam locomotive, the Britannia class Pacifics, arrived there in 1951, to take over the principal express passenger services. But, despite this influx of two important new locomotive types, the B17 Sandringham class was not completely ousted. But the next major development in operations to affect the class was the great Modernisation Programme of 1955, and it did bring about the final disappearance of the class, with only a handful left by the end of the decade.

Their final demise was not before some further examples had received Thompson's diagram 100A boiler. In the series of comparative trials held between one of the Sandringhams rebuilt to class B2 and an original B17, along with a B17 with the B1 type boiler, the latter had demonstrated great potential with its new steam plant. Influenced by the performance of the B17 carrying the 100A boiler, another fifty-five B17s were so fitted between 1947 and 1958, with their boilers pressed to work at 2251 lb/sq in, and were reclassified B17/6. As BR locomotives they were initially Class 4P, with a route availability of 5, which placed them in the same category as the B1, K2, and various 0-6-0 types, and even the 1,500V dc BoBo electric locomotives. An exception to this case was the streamlined locomotives of Class B17/5, which had a route availability of 7. From 1953 the B17s with the new diagram 100A boilers, classified B17/6, had had their power rating upgraded to 5P.

B17 Allocations at December 1959
31A Cambridge
  61607/8/10/13–16/18/23/25/26/32/39/44/
  47/51/52/55/61/62 (20)
31B March
  61627/33/35/41/45/46/53/56/57/60/64 (11)
31C Kings Lynn
  61620 (1)
30A Stratford
  61658/63/66/68 (4)
32A Norwich
  61611/36/54 (3)
32B Ipswich
  61612/29/31/37/49 (5)
32C Lowestoft
  61659/70/72 (3)
32D Yarmouth (South Town)
  61665 (1)

By the spring of 1960, only 17 locomotives remained in service, the majority of which were at the Cambridge and March depots, with three each at Stratford and Lowestoft. By the end of the year all had gone, with not a single example rescued for preservation. The last member of the class to go was No 61668 which had been withdrawn in August 1960 and stored at Southend for a month before being sent to Stratford Works for scrapping. Before its withdrawal it was used occasionally, and one of these last duties had involved rescuing a failed air-braked electric multiple unit – No 61668 was the only air-braked motive power available!

THE THOMPSON B1 CLASS 4-6-0

During the course of a successful career, there were no major rebuilds or modifications to this class although an ongoing problem of plate fractures in the firebox did give rise to the suggestion that a BR Standard No 3 boiler be fitted.

The early years under new ownership included participation in the Interchange Trials of 1948, and an exhaustive series of performance and efficiency tests carried out by the Railway Executive in 1950/51 with No 61353. This locomotive had been sent as built, albeit with some minor adaptations especially for the trials, to Rugby Testing Station in the autumn of 1949. The tests did not begin in earnest until more than a year later, in late 1950, with the results and some conclusions published in August 1951 in the form of a British Transport Commission test bulletin.

Construction and Operations

The first noticeable change to occur after 1948 was the addition of a temporary running number prefix E to denote regional ownership, until a full renumbering scheme was decided upon and announced later in that year. The B1s, like other Eastern and North Eastern Region locomotives, acquired 60,000 to their post-1946 LNER running numbers, which then ultimately ran from 61000 to 61409. The first locomotive to emerge from works with a BR running number was 61304, while the last to receive its new five-figure number was 61038 in November 1950. Prior to the introduction of 60,000 numbers on the Eastern and North Eastern Regions, 16 of the class were out-

shopped with the E prefix when new, Nos 61288 to 61303, along with a number of other members of the class, during these first few months of national ownership.

The largest allocations at the following depots at the end of 1947 were:

34A King's Cross	14
34D Hitchin	14
35A New England	10
36A Doncaster	19
38C Leicester (GC)	12
39A Gorton	15
30A Stratford	11
32A Norwich	17
50B Leeds (Neville Hill)	12

By the early 1950s, B1s were arriving at many other depots in considerable numbers. Immingham and Lincoln had seen their allocations rise to 22 and 11 respectively, from just six and one in 1947. Changes were made to allocations in the Eastern Region, and numbers rose in Scotland. In East London the principal depot Stratford, saw its share of the class rise to 29 by 1952, and it remained a dominant location throughout the 1950s.

Just four years after nationalisation, the whole class had been delivered, and throughout the Eastern and North Eastern Regions of BR only the Tyne/Tees area lacked any major allocation for a time, although Stockton and Darlington depots were housing 12 and 18 locomotives respectively in 1952. At the same time north of the border, 70 of the class had been sent to work, with the largest concentrations at Eastfield (Glasgow) and Kittybrewster (Aberdeen). Former Great Central sheds like Sheffield (Darnall), Gorton and Immingham housed 55 of the class between them, with the GE Section acquiring another 43 at Parkeston, Ipswich, Norwich, and Cambridge.

From the final orders placed by the LNER, which appeared between November 1948 and October the following year, four ex-GCR B8s were withdrawn, providing numbers 1353/5/7/8 for the new locomotives. The first BR order for B1s demanded the renumbering of yet more former GCR types. Eleven of Class B7 were given new numbers to make way for the incoming B1s. It was also necessary in 1949 to renumber some B16 4-6-0s to clear Nos 61400–61409, which became the final batch of 10 B1s built under the February 1949 order.

Delivery of the final batches began in 1950 and by April 1952, due to delays in the completion of the NBL contract, the total of 409 B1s was distributed as follows:

Eastern Region (Western Section)	187
Eastern Region (Eastern Section)	72
North Eastern Region	80
Scottish Region	70
Total	409

By this time, the class was widely distributed throughout the former LNER system, and throughout their careers, in addition to re-allocation within areas or regions there was some inter-regional transfer. Much of this took place as BR reorganised its regions on a more geographical basis in some instances, and changes in regional boundaries resulted in the transfer of former LNER depots and routes to London Midland Region management. Some depots in the West Yorkshire area were transferred from Eastern to North Eastern Region control, while Carlisle Canal depot, which had been transferred to the Scottish Region, was returned to the LM Region in the 1950s along with its B1s. LM Region had also acquired a number of former GCR sheds, such as Gorton, Annesley, and Leicester, also with B1 4-6-0s.

An example of inter-regional transfer took place in 1953, when 14 B1s were sent to the Southern Region in May as substitute motive power when Bulleid Pacifics were taken out of service in significant numbers for attention to their coupled axles. All the B1s transferred to the Southern at this time were stabled at Stewarts Lane, and used on passenger services between London Victoria and Dover or Ramsgate. The visit to the Southern in such numbers and on a regular basis was short-lived, some lasting only for a few days, with all returning to their home depots by June.

The GE Section which had been the first regular user of B1s was also one of the first areas to be fully dieselised, and on the 9 September 1962 No 61156 worked the last steam-hauled passenger train into Liverpool Street station. Even after this there was some work for the B1s on the GE Section, on a regular Wisbech to Whitemoor goods train in 1963. In that same year, B1s based at March were to be found working freight trains. March was the last shed to be closed to

steam in the area, in December 1963.

In the North East, there was much re-allocation of locomotives throughout the 1950s, transferring between districts and regions, as organisational changes were made. This constant changing of home bases was emphasised as new diesel locomotives began to appear from the late 1950s, although wholesale withdrawal of the class in the North East did not begin until the 1962–1965 period. By that time there were fifty-six B1s still in service in the area, with the majority housed at Ardsley, Wakefield and York. Except for the B1s transferred to service stock, Nos 61030, 61306 and 61337 were the last of the class in regular use by BR, being withdrawn in September 1967.

Locomotives allocated to depots on the former Great Northern and Great Central lines saw some work on the most prestigious as well as the humbler duties. On the latter routes they had been found at the head of such trains as the Master Cutler and South Yorkshireman, but were subsequently replaced by an influx of Pacific locomotives. During the mid-1950s, the Pacifics returned to their home regions and the East Coast Main Line, and the B1s assumed greater control over the whole length of former GCR routes. They also worked further afield, especially following the transfer of some depots to the LMR, on summer excursions as far as Bristol, and on a regular, Saturday Sheffield to Llandudno service.

In Scotland, after all the class had been delivered to the North British and GNofS sections, they had taken over passenger duties formerly worked by older 4-4-0 types, and covered ultimately all the main lines out of Edinburgh, and some secondary routes. During the 1950s Scottish locomotives could be found on the humbler duties, even acting as station pilots on occasions.

During the 1960s, no fewer than 17 of the class were transferred to service stock, nine of these in November 1963, and the remainder on various dates between January 1965 and February 1966. All were withdrawn from running stock, acquiring new departmental numbers, although No 61323 as Departmental No 24 was withdrawn almost immediately on its transfer. As departmental stock, they were used as stationary boilers at locations on the Eastern Region, with their drawgear and coupling rods removed.

Class B1 4-6-0s transferred to departmental stock

Running No	Transferred	Departmental number	Withdrawn
61059	11/63	17	4/66
61181	11/63	18	12/65
61204	11/63	19	2/66
61205	11/63	20	1/65
61233	11/63	21	4/66
61252	11/63	22	5/64
61300	11/63	23	11/65
61323*	11/63	24	11/63
61375	11/63	24	4/66
61272	1/65	25	11/65
61138	1/65	26	10/67
61105	3/65	27	5/66
61194	8/65	28	6/66
61264	11/65	29	7/67
61050	2/66	30	4/68
61051	2/66	31	3/66
61315	2/66	32	4/68

* This locomotive was scrapped soon after transfer to service stock.

The withdrawal of B1 class 4-6-0s began towards the end of 1961, when some of the class were barely 11 years old. Over the following five years 319 had been taken out of stock, just under 64 locomotives a year, or five a week, leaving only 90 in service in 1966. The following year, the last of regular revenue-earning service for the class, saw only 34 left, at Hull (Dairycoates), Low Moor, Wakefield, York, Dundee (Tay Bridge), Dunfermline, and Thornton Junction. The last to be scrapped were Departmental Nos 30 and 32, in April 1968.

Nos 61264 and 61306 have been rescued for preservation. No 61306 has been used regularly on BR metals for a number of years, repainted in LNER livery, and carrying the name *Mayflower*, as LNER No 1306. This name was carried previously by No 61379.

Interchange trials and performance testing
The B1 class 4-6-0 was selected to represent the design practices of the LNER, as a mixed-traffic type, in the series of trials organised by British Railways in 1948. Nos 61163, 61251/92 were chosen, and they were put up against the GWR Modified Hall, SR West Country Pacific, and the LMSR Class 5. The testing period ran from 31 May to 19 July 1948, with four routes earmarked for testing the mixed-traffic locomotives: Marylebone to Manchester, St Pancras to Man-

chester, Bristol to Plymouth, and Perth to Inverness. In each case, the routes provided severe tests of engine power, as indeed they would have in normal service, with some notoriously steep gradients. The first runs which included the B1 took place in the week beginning 31 May with the 10.00am Marylebone to Manchester, while the West Country Pacific crew was learning the route. The service trains used for the purposes of the trials on other regions were: London Midland, 10.15am St Pancras to Manchester; Western, the 1.45pm Bristol to Plymouth; in Scotland the 4.00pm Perth to Inverness.

While it was not an overriding parameter of the trials that the locomotives which performed best or most economically would provide the basis of future designs, the B1s put up some interesting runs. On home metals, former GCR lines, the locomotive was out-performed in the economy stakes by the ex-LMS Class 5. Similar results, if not worse, were achieved away from home, where on the Western Region, both the Stanier and Thompson 4-6-0s had the same disadvantages, the B1 managed to consume about 20% more fuel by comparison. On the former Midland lines, some curious mechanical problems afflicted No 61251, where at short cut-offs a knocking effect was produced in the trailing axleboxes, with a marked side-to-side movement when running at high speed and short cut-offs, or coasting. The only other mechanical troubles to affect the running was the self-cleaning apparatus which, becoming clogged with particles of ash, was impairing the draughting and steaming of the boiler. The equipment was removed for the remainder of the tests, and was done before the locomotive was sent for trials on the former Highland Line in Scotland.

Performance and efficiency tests with No 61353
These extensive tests were conducted during late 1950 at Rugby on the stationary test plant, and in 1951 over the route between Skipton and Carlisle. As built No 61353 was not fitted with self-cleaning apparatus, and although some preliminary changes were made varying the dimensions of blastpipe and chimney the standard arrangement was retained for the tests. This layout, with $5\frac{1}{8}$ in diameter blastpipe, 15 in diameter chimney choke, and an internal taper of 1 in 26, was

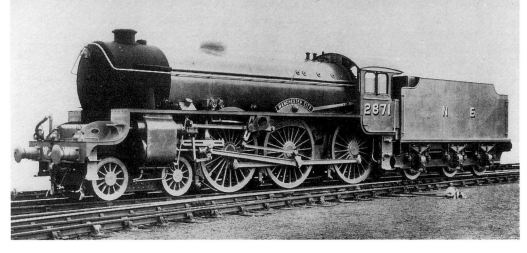

The first of the Thompson rebuilds of the Sandringhams, No 2871 *Manchester City*, with the diagram 100A boiler, two outside cylinders, and two sets of outside valve gear. Evidence of wartime economy can be seen in the simple 'N E' logo on the tender. *L&GRP*

sufficient to allow the locomotive to steam up to the grate limit.

The locomotive had run 46,000 miles since being built in 1949, and preparatory work for the tests included re-profiling the tyres, re-conditioning axleboxes, and the fitting of new rings to pistons and valves. In order to minimise the excessive reciprocating forces on the drawbar of the Rugby plant, the loco-motive was re-balanced so that 70% of the reciprocating masses were balanced, com-pared with the normal 30%.

Under test at Rugby, feedwater rates varied from 7,700 to 20,200 lb/hr with the exhaust steam injector in use, and up to 25,000 lb/hr with the live steam injector only. Cut-off positions varied between 15% and 45%, at nominal speeds from 15 mph to 70 mph. Coal consumption tests occupied periods of between 80 and 160 minutes, ac-cording to the firing rate. Swindon methods of testing were used, based on constant steam rates – the rates were kept constant out on the road with the dynamometer car, by adjusting the cut-off to suit the train speed. The cut-off was set by reference to a special steam flow meter developed by Swindon, and calibrated for the individual locomotive during the stationary test.

Out on the line, running with Dynamom-eter Car No 1, train loads varied from 116 to 436 tons, depending on the steam rate chosen for the particular test. Running from Carlisle to Skipton, with long stretches at 1 in 100, these tests took about an hour, keeping to prevailing passenger train schedules, but were of only 35 minutes' duration in the

opposite direction, due to a temporary speed restriction. While the locomotive rode well, and its boiler proved very free steaming, the report of the tests published in August 1951 suggests that it did not respond quickly enough in recovering from any departure from correct methods or rates of firing. The fire was kept thin but even with the firehole door closed when not firing, with the front ashpan damper open one notch at low power rates, and two at higher powers. The reports conclude with graphs of performance data obtained on both the stationary and con-trolled road tests, and summarising on the nature and their validity to the compilation of operating schedules, rather than the oper-ation and performance of the B1 design itself. No comparative figures are given, for instance, showing the design compared with other similar types, although to be fair BR was embarking at that stage on a long-term testing programme.

These were the most comprehensive trials with an LNER 4-6-0, but lesser trials were conducted in 1951 with three more B1s. In March and April for instance, Nos 61370/3 worked double-headed trains, hauling the dynamometer car, over a level route between Grimsby and New England. These constant speed tests, during which wind speed and direction indicators were carried fixed to the front of the locomotive in addition to the normal test equipment, were repeated be-tween Whitemoor and Norwich with freight trains. Lastly in June 1951 some small com-parative trials were made on the East Anglian service, alongside a Britannia class 4-6-2 be-tween Liverpool Street and Norwich. The BR Standard Pacifics were new and bore some similarity with the B1 class, since they were both two-cylinder mixed-traffic types, running on 6 ft 2 in coupled wheels.

ALLOCATIONS OF LNER
4-6-0 CLASSES – APRIL 1937

B1 (B18) Class:	Woodford:	5195/6
B2 (B19) Class:	Gorton:	5427
	Immingham:	5423
	Lincoln:	5424
	Sheffield:	5425/6/8
B3 Class:	Immingham:	6164/7/9
	Neasden:	6165/6/8
B4 Class:	Copley Hill:	6096–9, 6100–2
	Lincoln:	6103–4
B5 Class:	Immingham:	6070/2
	Sheffield:	6067–9
	Woodford:	6071
B6 Class	Sheffield:	5052/3, 5416
B7 Class:	Annesley:	5032, 5469, 5471/2
	Colwick:	5470
	Gorton:	5031/2/4–7/78, 5458/75
	Immingham:	5467/76–8
	Neasden:	5459/62/6/8/73/4
	Sheffield:	5033, 5480/3
	Woodford:	5038/72/3, 5460/1/3–5/79/81/2/4
B8 Class:	Annesley:	5279/80, 5440/2/6
	Colwick:	5004, 5439/41/3–5
B9 Class:	Gorton:	6105/6/9/11–14
	Trafford Park:	6107/8/10
B12 Class:	Cambridge:	8511/2/4/20–2/7/30
	Colchester:	8513/8/33/38/67/76/8/80
	Elgin:	8502/39
	GNofS:	8526
	Ipswich:	8515/9/25/9/35/7/44/62–5/77
	Kittybrewster:	8500/1/3/4/24/8/31/6/48
	Norwich:	8540/66/9/70
	Parkeston:	8507/8/32/68
	Stratford:	8505/9/16/17/41–3/5–7/49–61/71–5/9
	Yarmouth:	8510/23/34
B13 Class:	Darlington:	761 (Service Stock)
	Hull (Botanic Gardens):	753
	Hull (Dairycoates):	738/48
	Hull (Springhead):	782
	Leeds:	762
	York:	759
B15 Class:	Heaton:	795/6/8, 825
	Hull (Dairycoates):	788, 815/23
	Hull (Springhead):	786
	Leeds:	819/24
	Selby:	797
	Starbeck:	791
	York:	787/99, 813/7/20–2

B16 Class:	Blaydon:	932
	Darlington:	915, 1381/2
	Gateshead:	914/20
	Heaton:	843/6
	Hull (Dairycoates):	840–2, 906/9/22/5/8/30/3/4/6/7/42, 1371/3–6/8/84/5
	Leeds:	848/9, 924/9/31, 1379/80/3
	Scarborough:	845
	Tayport:	943
	York:	844/7, 908/11/21/3/6/7, 1372/7
B17 Class:	Cambridge:	2817–9
	Doncaster:	2832/3/5
	Gorton:	2816/24/34/40–2/59–62
	Ipswich:	2806/20/5/45
	Leicester:	2848–55
	March:	2821/9/46
	Norwich:	2837–9/43/4
	Neasden:	2847/56/7
	Parkeston:	2809/4/5/22/3/6/7/31/6
	Stratford:	2800/2/3/8–15/28/30/58
	Sheffield:	2863–5
	Woodford:	2866

In almost as new condition No 1060 of Edward Thompson's B1 class stands on shed, wearing LNER black livery. The cantilever brackets supporting the running boards can be clearly seen in this view. *Lens of Sutton*

INDEX